T0261787

POPULATION DEMOGRAPHY *of* NORTHERN SPOTTED OWLS

STUDIES IN AVIAN BIOLOGY

A Publication of the Cooper Ornithological Society

WWW.UCPRESS.EDU/GO/SAB

Studies in Avian Biology is a series of works published by the Cooper Ornithological Society since 1978. Volumes in the series address current topics in ornithology and can be organized as monographs or multi-authored collections of chapters. Authors are invited to contact the series editor to discuss project proposals and guidelines for preparation of manuscripts.

Series Editor
Brett K. Sandercock, Kansas State University

Editorial Board
Frank R. Moore, University of Southern Mississippi
John T. Rotenberry, University of California at Riverside
Steven R. Beissinger, University of California at Berkeley
Katie M. Dugger, Oregon State University
Amanda D. Rodewald, Ohio State University
Jeffrey F. Kelly, University of Oklahoma

Science Publisher
Charles R. Crumly, University of California Press

See complete series list on page 105.

POPULATION DEMOGRAPHY *of* NORTHERN SPOTTED OWLS

Eric D. Forsman, Robert G. Anthony, Katie M. Dugger,
Elizabeth M. Glenn, Alan B. Franklin, Gary C. White,
Carl J. Schwarz, Kenneth P. Burnham, David R. Anderson,
James D. Nichols, James E. Hines, Joseph B. Lint,
Raymond J. Davis, Steven H. Ackers, Lawrence S. Andrews,
Brian L. Biswell, Peter C. Carlson, Lowell V. Diller,
Scott A. Gremel, Dale R. Herter, J. Mark Higley,
Robert B. Horn, Janice A. Reid, Jeremy Rockweit,
James P. Schaberl, Thomas J. Snetsinger, and Stan G. Sovern

Studies in Avian Biology No. 40

A PUBLICATION OF THE COOPER ORNITHOLOGICAL SOCIETY

University of California Press

Berkeley   Los Angeles   London

University of California Press, one of the most distinguished university presses in the United States, enriches lives around the world by advancing scholarship in the humanities, social sciences, and natural sciences. Its activities are supported by the UC Press Foundation and by philanthropic contributions from individuals and institutions. For more information, visit www.ucpress.edu.

Studies in Avian Biology, No. 40
For digital edition of this work, please see the UC Press website.

University of California Press
Berkeley and Los Angeles, California

University of California Press, Ltd.
London, England

© 2011 by the Cooper Ornithological Society

Library of Congress Cataloging-in-Publication Data

Population demography of northern spotted owls / Eric D. Forsman ... [et al.].
p. cm. — (Studies in avian biology ; no. 40)
"A Publication of the Cooper Ornithological Society."
Includes bibliographical references and index.
ISBN 978-0-520-27008-4 (cloth : alk. paper)
1. Northern spotted owl.   2. Bird populations. I. Forsman, Eric D.

QL696.S83P665 2011
598.9'7--dc22   2010048821

19   18   17   16   15   14   13   12   11
10   9   8   7   6   5   4   3   2   1

The paper used in this publication meets the minimum requirements of
ANSI/NISO Z39.48-1992 (R 1997)(*Permanence of Paper*).

Cover photo: Adult female Spotted Owl in the Oregon Coast Range. Photo by Patrick Kolar.

PERMISSION TO COPY

The Cooper Ornithological Society hereby grants permission to copy chapters
(in whole or in part) appearing in *Studies in Avian Biology* for personal use,
or educational use within one's home institution, without payment,
provided that the copied material bears the statement
"© 2011 The Cooper Ornithological Society"
and the full citation, including names of all authors and year of publication.
Authors may post copies of contributed chapters on personal or institutional websites,
with the exception that complete volumes of *Studies in Avian Biology* may not be
posted on websites. Any use not specifically granted here, and any use of
Studies in Avian Biology articles or portions thereof for advertising, republication,
or commercial uses, requires prior consent from the series editor.

CONTENTS

AUTHORS

STEVEN H. ACKERS
Oregon Cooperative Fish and Wildlife
Research Unit
Department of Fisheries and Wildlife
Oregon State University
Corvallis, OR 97331
ackerss@onid.orst.edu

LAWRENCE S. ANDREWS
Oregon Cooperative Fish and Wildlife
Research Unit
Department of Fisheries and Wildlife
Oregon State University
Corvallis, OR 97331
sandrewsor@aol.com

DAVID R. ANDERSON
Department of Fish, Wildlife, and
Conservation Biology
Colorado State University
Fort Collins, CO 80523
quietanderson@yahoo.com

ROBERT G. ANTHONY
U.S. Geological Survey
Oregon Cooperative Fish and Wildlife
Research Unit
Department of Fisheries and Wildlife
Oregon State University
Corvallis, OR 97331
robert.anthony@oregonstate.edu

BRIAN L. BISWELL
USDA Forest Service
Pacific Northwest Research Station
Olympia Forestry Sciences Lab
3625 93rd Avenue SW
Olympia, WA 98512
bbiswell@fs.fed.us

KENNETH P. BURNHAM
U.S. Geological Survey
Colorado Cooperative Fish and Wildlife
Research Unit, Department of Fish,
Wildlife, and Conservation Biology
Colorado State University
Fort Collins, CO 80523
kenb@lamar.colostate.edu

PETER C. CARLSON
Department of Fish, Wildlife, and
Conservation Biology
Colorado State University
Fort Collins, CO 80523
pccarlson@lamar.colostate.edu

RAYMOND J. DAVIS
USDA Forest Service
Umpqua National Forest
2900 Stewart Parkway
Roseburg, OR 97471
rjdavis@fs.fed.us

LOWELL V. DILLER
Green Diamond Resource Company
900 Riverside Road
Korbel, CA 95550
ldiller@greendiamond.com

KATIE M. DUGGER
Department of Fisheries and Wildlife
Oregon State University
Corvallis, OR 97331
katie.dugger@oregonstate.edu

ERIC D. FORSMAN
USDA Forest Service
Pacific Northwest Research Station
Corvallis Forestry Sciences Lab
3200 Jefferson Way
Corvallis, OR 97331
eforsman@fs.fed.us

ALAN B. FRANKLIN
USDA/APHIS
National Wildlife Research Center
4101 LaPorte Avenue
Fort Collins, CO 89521
alan.b.franklin@aphis.usda.gov

ELIZABETH M. GLENN
Oregon Cooperative Fish and Wildlife
Research Unit
Department of Fisheries and Wildlife
Oregon State University
Corvallis, OR 97331
betsyglenn1@gmail.com

SCOTT A. GREMEL
USDI National Park Service
Olympic National Park
600 East Park Avenue
Port Angeles, WA 98362
scott_gremel@nps.gov

DALE R. HERTER
Raedeke Associates, Inc.
5711 NE 63rd Street
Seattle, WA 98115
drherter@raedeke.com

J. MARK HIGLEY
Hoopa Tribal Forestry
P.O. Box 368
Hoopa, CA 95546
mhigley@hoopa-nsn.gov

JAMES E. HINES
U.S. Geological Survey
Patuxent Wildlife Research Center
12100 Beech Forest Road
Laurel, MD 20708
jhines@usgs.gov

ROBERT B. HORN
USDI Bureau of Land Management
Roseburg District Office
777 Garden Valley Boulevard
Roseburg, OR 97471
robert_horn@or.blm.gov

JOSEPH B. LINT
USDI Bureau of Land Management
Roseburg District Office
777 Garden Valley Boulevard
Roseburg, OR 97471
jbralint@msn.com

JAMES D. NICHOLS
U.S. Geological Survey
Patuxent Wildlife Research Center
12100 Beech Forest Road
Laurel, MD 20708
jnichols@usgs.gov

JANICE A. REID
USDA Forest Service
Pacific Northwest Research Station
Roseburg Field Station
777 Garden Valley Boulevard
Roseburg, OR 97471
j1reid@or.blm.gov

JEREMY ROCKWEIT
Department of Fish, Wildlife, and
Conservation Biology
Colorado State University
Fort Collins, CO 80523
rockweit@rams.colostate.edu

JAMES P. SCHABERL
Mt. Rainier National Park
Tahoma Woods Star Route
Ashford WA 98304
jim_schaberl@nps.gov

CARL J. SCHWARZ
Department of Statistics and Actuarial Science
Simon Fraser University
8888 University Drive
Burnaby, BC, V5A 1S6, Canada
cschwarz@sfu.ca

THOMAS J. SNETSINGER
Oregon Cooperative Fish and Wildlife
Research Unit
Department of Fisheries and Wildlife
Oregon State University
Corvallis, OR 97331
tom.snetsinger@oregonstate.edu

STAN G. SOVERN
Oregon Cooperative Fish and Wildlife
Research Unit
Department of Fisheries and Wildlife
Oregon State University
Corvallis, OR 97331
ssovern@fs.fed.us

GARY C. WHITE
Department of Fish, Wildlife, and
Conservation Biology
Colorado State University
Fort Collins, CO 80523
gwhite@cnr.colostate.edu

ACKNOWLEDGMENTS

We are particularly indebted to Rich Fredrickson, Rocky Gutiérrez, Patti Happe, Pete Loschl, Chuck Meslow, Gary Miller, Bruce Moorhead, Frank Oliver, Len Ruggerio, Erran Seaman, and Pat Ward, who were instrumental in the initiation of one or more of the 11 demographic studies of Northern Spotted Owls. Frank Oliver bent, drilled, and polished over 10,000 color bands that we used to mark Spotted Owls. This study would not have been possible without the assistance of a small army of dedicated field biologists who helped collect the data. There have been so many participants that we cannot remember all the names, but the ones who contributed the most were: Steve Adams, Steve Adey, Garth Alling, Meg Amos, David Anderson, Scott Armentrout, Duane Aubuchon, Keith Bagnall, Doug Barrett, Greg Bennett, Ryan Besser, Jennifer Blakesley, Wendi Bonds, John Bottelli, Robin Bown, Cheryl Broyles, Mathew Brunner, Erin Burke, Tim Burnett, Joe Burns, Patty Buettner, Sue Canniff, Adam Canter, Chris Cantway, Bruce Casler, Doug Chamblin, Rita Claremont, David Clarkson, Billy Colegrove, Alan Comulada, Caroline Crowley, Ron Crutchley, Sam Cuenca, Natalie Cull, Amy Cunkelman, David Delaney, Glenn Desy, Diane Doersch, Michelle Dragoo, Amy Ellingson, Roli Espinosa, Thomas Evans, Annie Farris, Mark Fasching, Tim Fetz, Helen Fitting, David Fix, Tracy Ford, Ray Forson, Gila Fox, Tim Fox, Shawna Franklin, Laura Friar, Ken Fukuda, Rudy Galindo, Dawn Garcia, Daniel George, Alan Giese, David Giessler, Charles Goddard, Teresa Godfrey, Jim Goode, Scott Graham, Eric Grant, Ashley Green, Allison Greenleaf, Tim Grubert, Gillian Hadley, Terry Hall, Tammy Hamer, Andy Hamilton, Keith Hamm, Timothy Hanks, Rick Hardy, Megan Harrigan, Steve Hayner, Amanda Heffner, Thomas Heinz, Marisa Herrera, Lorin Hicks, Connie Holloway, Taryn Hoover, Eric Hopson, Scott Horton, Gay Hunter, John Hunter, Mark Irwin, Eugene Jackson, Ryan Jackson, Colin Jewett, Aaron Johnston, Jennifer Jones, Leilani Jones, Timm Kaminski, Marc Kasper, Tom Kay, Debaran Kelso, Leah Kenney, Wendy King, Wes King, Charles Knight, Mike Koranda, Richard Kosteke, Stan Kott, Amy Krause, David Lamphear, Chris Larson, Megan Larsson, Kurt Laubenheimer, Richard Leach, Rich Lechleitner, Doug Leslie, Krista Lewiciki, Jeff Lewis, Patrick Loafman, Dan Loughman, Joanne Lowden, Rich Lowell, Hunter Lowry, Jessica Lux, Devin Malkin, Dave Manson, Kevin Maurice, Chris McCafferty, Aaron McKarley, Dawn McCovey, Andrew McLain, Brian Meiering, Merri Melde, Joel Merriman, William Meyer, Christine Moen, Jason Mowdy, Ellen Meyers, Ben Murphy, Michael Murray, Suzanne Nelson, Matt Nixon,

Timothy Parker, Heidi Pederson, Kim Pederson, Nicole Nielson-Pincus, Melanie O'Hara, Mary Oleri, Eduardo Olmeda, Michell O'Malley, Ivy Otto, Rob Owens, Antonio Padilla, Lyle Page, David Pavlackey, John Perkins, Jim Petterson, Mark Phillip, Tom Phillips, Tim Plawman, Aaron Pole, Andrew Pontius, Clayton Pope, Amy Price, Laura Quattrini, Laura Ratti, Elise Raymond, Mary Rasmussen, Reba Reidner, Angela Rex, Gail Rible, Kurt Richardson, Melanie Roan, Ana Roberts, Susan Roberts, Rob Roninger, Trish Roninger, Jen Sanborn, Ted Schattenkerk, Anthony Scheiff, Britta Schielke, Jason Schilling, Matthew Schlesinger, Kary Schlick, Doreen Schmidt, Michael Schwerdt, Tim Selim, Shawn Servoss, Paula Shaklee, Erica Sisson, Gail Sitter, Steve Small, Denise Smith, Don Smith, Spencer Smith, Alexis Smoluk, Maurice Sommé, Noel Soucy, Susannah Spock, Amy Stabins, Peter Steele, Cynthia Stern, Lori Stonum, Michael Storms, Bob Straub, Denise Strejc, Keith Swindle, Jim Swingle, Margy Taylor, Campbell Thompson, Joel Thompson, Jim Thrailkill, Sheila Turner-Hane, Shem Unger, Jennifer Van Gelder, Andrew Van Lanen, Nick Van Lanen, Veronica Vega, Frank F. Wagner, Fred Wahl, Stephanie Waldo, Abe Walston, Fred Way, Kevin White, Mark Williams, Kari Williamson, Heather Wise, Jim Woodford, Sarah Wyshynski, Lance Wyss, Bryan Yost, Kendall Young, Jessica Zeldt, and Joe Zisa. We also thank all the data analysts who helped analyze the data at the workshop, including Simon Bonner, Wendell Challenger, Mary M. Conner, Greg Davidson, Paul Doherty, Kate Huyvaert, Jake Ivan, David Lamphear, Julien Martin, Brett McClintock, and Trent McDonald. We could not have asked for a better group to get us through this difficult task in one week. Reviews of the manuscript by John Marzluff, E. Chuck Meslow, Martin Raphael, John Laurence, and two anonymous reviewers were extremely helpful. Jane Toliver at the Oregon Cooperative Wildlife Research Unit provided administrative support for the workshop. Funding for demographic studies of Northern Spotted Owls on federal lands was provided primarily by the USDA Forest Service, USDI Bureau of Land Management, and USDI National Park Service. Funding for studies on non-federal lands came from the Green Diamond Timber Company, Plum Creek Timber Company, Louisiana Pacific Lumber Company, and the Hoopa Tribe. Funding for the workshop was provided by the USDA Forest Service, USDI Bureau of Land Management, and USDI Fish and Wildlife Service.

Population Demography of Northern Spotted Owls

Abstract. We used data from 11 long-term studies to assess temporal and spatial patterns in
fecundity, apparent survival, recruitment, and
annual finite rate of population change of
Northern Spotted Owls (*Strix occidentalis caurina*) from 1985 to 2008. Our objectives were to
evaluate the status and trends of the subspecies
throughout its range and to investigate associations between population parameters and covariates that might be influencing any observed
trends. We examined associations between population parameters and temporal, spatial, and
ecological covariates by developing a set of *a
priori* hypotheses and models for each analysis.
We used information-theoretic methods and
$QAIC_c$ model selection to choose the best
model(s) and rank the rest. Variables included
in models were gender, age, and effects of time.
Covariates included in some analyses were
reproductive success, presence of Barred Owls
(*Strix varia*), percent cover of suitable owl habitat, several weather and climate variables including seasonal and annual variation in precipitation and temperature, and three long-term
climate indices. Estimates of fecundity, apparent
survival, recruitment, and annual rate of population change were computed from the best models or with model averaging for each study area.
The average number of years of reproductive data
from each study area was 19 (range = 17 to 24),
and the average number of captures/resightings
per study area was 2,219 (range = 583 to 3,777),
excluding multiple resightings of the same individuals in the same year. The total sample of 5,224
marked owls included 796 1-yr-old subadults, 903
2-yr-old subadults, and 3,545 adults (≥3 yrs old).
The total number of annual captures/recaptures/
resightings was 24,408, and the total number of
cases in which we determined the number of
young produced was 11,450.

Age had an important effect on fecundity,
with adult females generally having higher
fecundity than 1- or 2-yr-old females. Nine of the
11 study areas had an even−odd year effect on
fecundity in the best model or a competitive
model, with higher fecundity in even years.
Based on the best model that included a time
trend in fecundity, we concluded that fecundity
was declining on five areas, stable on three areas,
and increasing on three areas. Evidence for an
effect of Barred Owl presence on fecundity on
individual study areas was somewhat mixed.
The Barred Owl covariate was included in the
best model or a competitive model for five study
areas, but the relationship was negative for four
areas and positive for one area. At the other six
study areas, the association between fecundity
and the proportion of Spotted Owl territories in
which Barred Owls were detected was weak or
absent. The percent cover of suitable owl habitat

was in the top fecundity model for all study areas in Oregon, and in competitive models for two of the three study areas in Washington. In Oregon, all 95% confidence intervals on beta coefficients for the habitat covariate excluded zero, and on four of the five areas the relationship between the percent cover of suitable owl habitat and fecundity was positive, as predicted. However, contrary to our predictions, fecundity on one of the Oregon study areas (KLA) declined with increases in suitable habitat. On all three study areas in Washington, the beta estimates for the effects of habitat on fecundity had 95% confidence intervals that broadly overlapped zero, suggesting there was less evidence of a habitat effect on fecundity on those study areas. Habitat effects were not included in models for study areas in California, because we did not have a comparable habitat map for those areas. Weather covariates explained some of the variability in fecundity for five study areas, but the best weather covariate and the direction of the effect varied among areas. For example, there was evidence that fecundity was negatively associated with low temperatures and high amounts of precipitation during the early nesting season on three study areas but not on the other eight study areas.

The meta-analysis of fecundity for all study areas (no habitat covariates included) suggested that fecundity varied by time and was parallel across ecoregions or latitudinal gradients, with some weak evidence for a negative Barred Owl (BO) effect. However, the 95% confidence interval for the beta coefficient for the BO effect overlapped zero ($\hat{\beta}$ = -0.12, SE = 0.11, 95% CI = -0.31 to 0.07). The best models from the meta-analysis of fecundity for Washington and Oregon (habitat covariates included) included the effects of ecoregion and annual time plus weak effects of habitat and Barred Owls. However, the 95% confidence intervals for beta coefficients for the effects of Barred Owls and habitat overlapped zero ($\hat{\beta}_{BO}$ = -0.104, 95% CI = -0.369 to 0.151; $\hat{\beta}_{HAB1}$ = -0.469, 95% CI = -1.363 to 0.426). In both meta-analyses of fecundity, linear trends (T) in fecundity were not supported, nor were effects of land ownership, weather, or climate covariates. Average fecundity over all years was similar among ecoregions except for the Washington–Mixed-Conifer ecoregion, where mean fecundity was 1.7 to 2.0 times higher than in the other ecoregions.

In the analysis of apparent survival on individual study areas, recapture probabilities typically ranged from 0.70 to 0.90. Survival differed among age groups, with subadults, especially 1-yr-olds, having lower apparent survival than adults. There was strong support for declining adult survival on 10 of 11 study areas, and declines were most evident in Washington and northwest Oregon. There was also evidence that apparent survival was negatively associated with the presence of Barred Owls on six of the study areas. In the analyses of individual study areas, we found little evidence for differences in apparent survival between males and females, or for negative effects of reproduction on survival in the following year.

In the meta-analysis of apparent survival, the best model was a random effects model in which survival varied among study areas (g) and years (t), and recapture rates varied among study areas, sexes (s), and years. This model also included the random effects of study area and reproduction (R). The effect of reproduction was negative ($\hat{\beta}$ = -0.024), with a 95% confidence interval that barely overlapped zero (-0.049 to 0.001). Several random effects models were competitive, including a second-best model that included the Barred Owl (BO) covariate. The estimated regression coefficient for the BO covariate was negative ($\hat{\beta}$ = -0.086), with a 95% confidence interval that did not overlap zero (-0.158 to -0.014). One competitive random effects model included a negative linear time trend on survival ($\hat{\beta}$ = -0.0016) with a 95% confidence interval (-0.0035 to 0.0003) that barely overlapped zero. Other random effects models that were competitive with the best model included climate effects (Pacific Decadal Oscillation, Southern Oscillation Index) or weather effects (early nesting season precipitation, early nesting season temperature). Ownership category, percent cover of suitable owl habitat, and latitude had little to no effect on apparent survival.

Apparent survival differed among ecoregions, but the ecoregion covariate explained little of the variation among study areas and years.

Estimates of the annual finite rate of population change (λ) were below 1.0 for all study areas, and there was strong evidence that populations on 7 of the 11 study areas declined during the study. For four study areas, the 95% confidence intervals for λ overlapped 1.0, so we could not conclude that those populations were declining. The weighted mean estimate of λ for all study areas was 0.971 (SE = 0.007, 95% CI = 0.960 to 0.983), indicating that the average rate of population decline in all study areas combined was 2.9% per year. Annual rates of decline were most precipitous on study areas in Washington and northern Oregon. Based on estimates of realized population change, populations on four study areas declined 40 to 60% during the study, and populations on three study areas declined 20 to 30%. Declines on the other four areas were less dramatic (5 to 15%), with 95% confidence intervals that broadly overlapped 1.0.

Based on the top-ranked *a priori* model in the meta-analysis of λ, there was evidence that ecoregions and the proportion of Spotted Owl territories with Barred Owl detections were important sources of variation for apparent survival (φ_t) and recruitment (f_t). There was some evidence that recruitment was higher on study areas dominated by federal lands compared to study areas that were on private lands or lands that included approximately equal amounts of federal and private lands. There also was evidence that recruitment was positively related to the proportion of the study area that was covered by suitable owl habitat.

We concluded that fecundity, apparent survival, and/or populations were declining on most study areas, and that increasing numbers of Barred Owls and loss of habitat were partly responsible for these declines. However, fecundity and survival showed considerable annual variation at all study areas, little of which was explained by the covariates that we used. Although our study areas were not randomly selected, we believe our results reflected conditions on federal lands and areas of mixed federal and private lands within the range of the Northern Spotted Owl because the study areas were (1) large, covering $\approx 9\%$ of the range of the subspecies; (2) distributed across a broad geographic region and within most of the geographic provinces occupied by the owl; and (3) the percent cover of owl habitat was similar between our study areas and the surrounding landscapes.

Key Words: Barred Owl, fecundity, Northern Spotted Owl, Northwest Forest Plan, population change, recruitment, *Strix occidentalis caurina*, *Strix varia*, survival

D uring the last 40 years, the management philosophy on federal forest lands in the United States has undergone profound changes as government agencies have become increasingly aware of the importance of federal lands in species conservation. Nowhere has this change been more controversial than in the Pacific Northwest (Washington, Oregon, and northern California), where attempts to maintain viable populations of Northern Spotted Owls (*Strix occidentalis caurina*), Marbled Murrelets (*Brachyramphus marmoratus*), red tree voles (*Arborimus longicaudus*), and other plants and animals that thrive in old forests have resulted in large reductions in harvest of old forests on federal lands (Ervin 1989, Durbin 1996). Because of the controversial nature of these changes and the need to know whether management policies were achieving desired objectives, the U.S. Forest Service and U.S. Bureau of Land Management initiated eight long-term mark–recapture studies of Northern Spotted Owls during 1985 to 1991 (Lint et al. 1999). The primary objective of these field studies was to provide federal agencies and the public with data on the status and trends of Spotted Owl populations and to determine if the management plans adopted by the agencies were resulting in recovery of the owl, which was listed as a threatened subspecies in 1990 (USDI Fish and Wildlife Service 1990). In addition, the recent invasion of Barred Owls (*Strix varia*) into the range of the Spotted Owl represents a competitive threat that many research groups are trying to assess. The information generated in these studies has been featured in many publications (Franklin 1992, Burnham et al. 1994, 1996, Forsman et al. 1996a, Franklin et al. 2000, Kelly et al. 2003, Hamer et al. 2007, Olson et al. 2004, 2005, Anthony et al. 2006, Bailey et al. 2009, Singleton 2010) and has played a key role in several court cases and in the development of the Northwest Forest Plan (NWFP). The NWFP is an interagency plan that was designed to protect all native plants and animals on federal lands within the range of the Northern Spotted Owl, while at the same time providing jobs and wood products (USDA Forest Service and USDI Bureau of Land Management 1994). The data from the long-term demography studies were also considered by the team that prepared the 2008 recovery plan for the Northern Spotted Owl (USDI Fish and Wildlife Service 2008) and by a committee of The Wildlife Society (2008) who commented on the plan. Research on the long-term demography of the Spotted Owl has focused attention on forest management and conservation of forest wildlife in the western United States. This research, and the controversy it has created, have changed forest management in the region and helped to bring about a general reassessment of the role of forest management in species conservation, forest ecosystem management, and human health (Thomas et al. 1993, USDA Forest Service and USDI Bureau of Land Management 1994, Dietrich 2003).

With any large-scale, long-term monitoring program, important criteria are consistency in methods and funding, and a consistent protocol for analyzing the data and reporting the results. Standard protocols are especially important in cases like the Spotted Owl, where (1) the economic stakes are high, (2) there is occasional disagreement regarding the potential for bias in the estimates of demographic parameters (Loehle et al. 2005, Franklin et al. 2006), and (3) where many different agencies and stakeholders are responsible for collecting the data. For the Northern Spotted Owl, the methods for collecting, analyzing, and reporting demographic data have been described by Franklin et al. (1996), Lint et al. (1999), Anderson et al. (1999), and Anthony et al. (2006). Because of considerable scientific and public interest in these studies, one of the key features in the monitoring program has been regularly scheduled workshops in which all of the researchers who are conducting demographic studies of Northern Spotted Owls, meet and conduct a meta-analysis of all of the demographic data (Lint et al. 1999). Since 1993, there have been four cooperative workshops, the results of which have been described in three published articles (Burnham et al. 1994, 1996, Anthony et al. 2006) and one unpublished report (Franklin et al. 1999). The most recent of these workshops was conducted in January 2009, where we completed an updated meta-analysis in which we analyzed all of the demographic data currently available on the Northern Spotted Owl, including an additional five years of data from 2004 to 2008, and modeled the demographic parameters as a function of a new set of environmental covariates. Our demographic analyses, which represent the most complete and up-to-date summary of the population status of the subspecies, are the focus of this volume of Studies in Avian Biology.

Estimates of vital rates and population trends are more interesting when there is some understanding of the environmental factors that may influence those estimates. Anthony et al. (2006) included covariates for the cost of reproduction and presence of Barred Owls in their analyses of survival and population trends of Spotted Owls, but they were not able to include habitat or weather covariates in their analysis. In our analysis, we included the same covariates examined by Anthony et al. (2006) but add several new range-wide weather covariates and habitat covariates in Washington and Oregon. Thus, our analysis is the most comprehensive to date in terms of the number of covariates examined. Our analysis also differed from earlier analyses of Spotted Owl populations (Burnham et al. 1994, 1996) in that we use the *f*-parameterization of Pradel's (1996) temporal symmetry model to estimate the annual rate of population change (λ), and examine trends in the components of population change, including survival and recruitment rates. Our analyses have led to some valuable insights regarding our ability to discern the possible influence of environmental covariates (e.g., habitat, Barred Owls, weather) on a species that has high temporal variation in survival and reproduction. Our general approach will be of interest to other research groups investigating population dynamics of other long-lived vertebrates with similar life histories.

Our purpose in this report is threefold. First, we wanted to determine if the declines in apparent survival and populations that were reported previously (Anthony et al. 2006) have continued or stabilized. Second, we used multiple covariates in the analysis of demographic rates in an attempt to better understand which environmental factors best explained annual and spatial variability in these rates. We reasoned that one or more of these covariates might explain the recent declines in demographic rates of the subspecies. Last, we report on the use of the *f*-parameterization of the Pradel (1996) temporal symmetry model to estimate components of the annual finite rate of population change (λ), including apparent survival and recruitment rates, one of the first applications of this new technique in demographic analyses of Northern Spotted Owl populations.

STUDY AREAS

We obtained data from 11 study areas, including three in Washington, five in Oregon, and three in California (Fig. 1). Study area names and acronyms used throughout the report are described in Table 1. Size of study areas ranged from 356 to 3,922 km² (Table 1). The total area covered by all 11 study areas (19,813 km²)

was equal to approximately 9% of the total range of the Northern Spotted Owl, which has been estimated at 230,690 km² (USDA Forest Service and USDI Bureau of Land Management 1994). Our study areas included one (GDR) that was entirely on private land, one (HUP) on an Indian Reservation, four (OLY, HJA, CAS, NWC) that were primarily on federal lands, and five (CLE, RAI, COA, TYE, KLA) that included a mixture of federal, private and state lands (Table 1). Of the 11 study areas, eight (OLY, CLE, COA, HJA, TYE, KLA, CAS, NWC) were established by the U.S. Forest Service and U.S. Bureau of Land Management to document the status of Northern Spotted Owls on federal lands within the region encompassed by the Northwest Forest Plan (Lint et al. 1999). In some analyses, we present results separately for these eight areas, which we refer to as "NWFP study areas" (Table 1, Appendix A). We made a distinction between types of study areas because the Northwest Forest Plan is the overarching interagency land management plan that applies to federal lands within the range of the Northern Spotted Owl, which is of special interest to federal land managers (USDA Forest Service and USDI Bureau of Land Management 1994).

Our study areas differed from those included in Anthony et al. (2006) in that data collection on three of the 14 study areas examined therein, was either discontinued or reduced, so data from those three areas (Wenatchee, Warm Springs, and Marin study areas) were no longer available for a meta-analysis. In addition, the OLY study area was reduced in size because of lack of funding, and the size of the GDR study area was expanded in 1998. In two cases (TYE, NWC), sizes of study areas in Table 1 are different than in Anthony et al. (2006), not because of any change in area, but because we mapped the boundaries based on boundaries used in analyses of population change. In contrast, the study area boundaries for the TYE and NWC study areas displayed in Anthony et al. (2006) included survey polygons in areas adjacent to the main study areas. Because of the changes in number and size of study areas and the addition of five years of data, results of

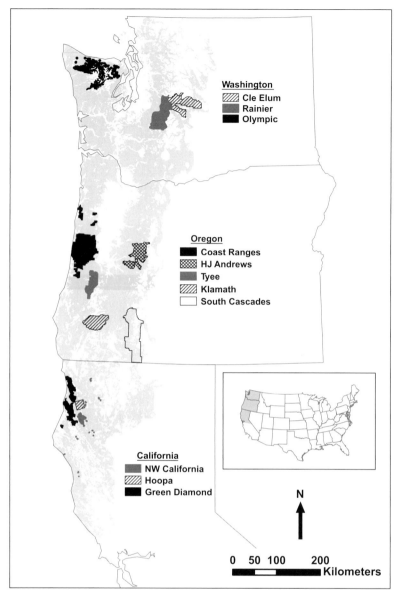

Figure 1. Locations of 11 study areas used in the analysis of vital rates and population trends of Northern Spotted Owls.

this analysis are not directly comparable to previous analyses conducted by Burnham et al. (1996) and Anthony et al. (2006).

The study areas were distributed across a broad geographic region, from central Washington south to northern California, and varied widely in climate, vegetation, and amount of topographic relief. Study areas in the coastal mountains of Oregon and California (COA, TYE, KLA, NWC, GDR,

HUP) typically occurred at low to moderate elevations, where the highest elevations were <1,250 m, whereas study areas in the Cascades and Olympic Mountains (CLE, RAI, OLY, HJA, CAS) occurred in areas with high mountains, where forests extended from the lowland valleys up to timberline, at or above 1,500 m elevation. Climate varied from relatively warm and dry on study areas in southern Oregon and northwestern California to

TABLE 1

Descriptions of 11 study areas used to estimate vital rates of Northern Spotted Owls in Washington, Oregon, and California (see also Appendix A).

Asterisks indicate the eight study areas that are part of the federal monitoring program for the northern spotted owl (Lint et al. 1999).

| Study area | Acronym | Years | Area (km^2) | No. owls banded by age class[a] | | | Total owls | Total encounters[b] | Mean annual precip. (cm) |
				S1	S2	Adults			
Washington									
Cle Elum[*]	CLE	1989–2008	1,784	31	32	148	211	1,170	142
Rainier	RAI	1992–2008	2,167	8	12	133	153	583	216
Olympic[*]	OLY	1990–2008	2,230	19	32	337	388	1,510	290
Oregon									
Coast Ranges[*]	COA	1990–2008	3,922	66	97	486	649	3,306	219
H. J. Andrews[*]	HJA	1988–2008	1,604	28	91	457	576	3,082	201
Tyee[*]	TYE	1990–2008	1,026	137	110	243	490	2,315	125
Klamath[*]	KLA	1990–2008	1,422	169	134	347	650	2,800	121
South Cascades[*]	CAS	1991–2008	3,377	43	80	479	602	2,364	123
California									
NW California[*]	NWC	1985–2008	460	114	80	280	474	2,550	155
Hoopa Tribe	HUP	1992–2008	356	38	47	130	215	951	195
Green Diamond	GDR	1990–2008	1,465	143	188	505	836	3,777	188
Totals			19,813	796	903	3,545	5,244	24,408	

[a] Age class codes indicate owls that were 1 year old (S1), 2 years old (S2), or ≥ 3 years old (Adults). Counts include owls first banded as S1's, S2's, or Adults, as well as owls first banded as juveniles that were subsequently recaptured when they were 1, 2, or ≥ 3 years old.
[b] All captures, recaptures, and re-sightings, excluding multiple encounters of individuals in the same year.

extremely wet in the temperate rain forests on the west side of the Olympic Peninsula, where annual precipitation ranged from 280 to 460 cm/year (Table 1). Study areas on the west slope of the Cascades Range (RAI, HJA, CAS) were typically warm and dry during summer and cool and wet during winter, with much of the winter precipitation falling as snow at higher elevations. The only study area that was entirely on the east slope of the Cascades (CLE) was characterized by warm, dry summers and cool winters, with most precipitation occurring as snow during winter.

Forests on all study areas were dominated by conifers, or mixtures of conifers and hardwoods,

but there were regional differences in species composition. Forests on study areas in Washington and northern Oregon were comprised of mixtures of Douglas-fir (*Pseudotsuga menziesii*) and western hemlock (*Tsuga heterophylla*), or, in coastal areas, by mixed stands of western hemlock and Sitka spruce (*Picea sitchensis*). Ponderosa pine (*Pinus ponderosa*) was also a dominant species on the east slope of the Cascades in Washington. Study areas in southwestern Oregon and northwestern California had diverse mixtures of mixed-conifer forest or mixed-evergreen forest (Franklin and Dyrness 1973, Küchler 1977). Common canopy trees in mixed-conifer or mixed-evergreen forests included: Douglas-fir, grand fir (*Abies grandis*), western white pine (*P. monticola*), sugar pine (*P. lambertiana*), ponderosa pine, incense cedar (*Calocedrus decurrens*), tanoak (*Lithocarpus densiflorus*), Pacific madrone (*Arbutus menziesii*), California laurel (*Umbellularia californica*), and canyon live-oak (*Quercus chrysolepis*). The GDR study area in coastal northwestern California also included considerable amounts of coast redwood (*Sequoia sempervirens*) forest at lower elevations.

Forest age and structure varied widely among areas, ranging from one study area (GDR) that was mostly dominated by forests that were <60 years old to some study areas on federal lands (OLY, HJA, NWC, CAS) in which >60% of the landscape was covered by mature (80 to 199 years old) and old-growth forests (≥200 years old) with multilayered canopies of large trees that were typically 50 to 200 cm diameter at breast height (dbh). All study areas were characterized by diverse mixtures of forest age classes that were the product of a long history of logging, fire, windstorms, disease, and insect damage. Forests on the OLY and RAI study areas were also naturally fragmented by high-elevation ridges that were covered by snowfields and bare rock.

As stated by Franklin et al. (1996) and Anthony et al. (2006), the 11 study areas in our analysis were selected based on many considerations, including forest type, logistics, funding, land ownership boundaries, and local support from management agencies. As a result, the study areas were not randomly selected or systematically spaced. However, the study areas covered ~9% of the range of the subspecies, and an analysis by Anthony et al. (2006) indicated that the amount of suitable owl habitat in the study areas was similar to the surrounding areas. We believe, therefore, that the habitat conditions within our study areas were broadly representative of conditions on federal lands within the range of the owl, and that our results are indicative of population attributes of Northern Spotted Owls on federal lands in general. We are less confident that our estimates reflect typical trends on non-federal lands because our sample

only included two study areas situated exclusively on non-federal lands (HUP and GDR). Both of those areas were in California, near the southern end of the range of the Northern Spotted Owl (Fig. 1) and were unique in that both landowners were actively managing to provide nesting and foraging habitat for Spotted Owls.

FIELD METHODS

We surveyed our study areas each year to locate owls, confirm bands, band unmarked owls, and document the number of young produced by each territorial female. Owls were trapped with a variety of methods, most commonly with a noose pole or snare pole (Forsman 1983). Each owl was marked with a U.S. Geological Survey numbered band on one leg and a unique color band on the other leg that could be observed without recapturing the owl (Forsman et al. 1996b, Reid et al. 1999). Surveys were conducted using vocal imitations or playback of owl calls to incite the owls to defend their territories, thereby revealing their presence (Franklin et al. 1996). However, once we became familiar with traditional nest and roost areas used by owls, it was often possible to locate owls by walking into traditional nest areas during the day and calling quietly while visually searching for owls near the nest. The number of visits to each survey polygon or owl territory within each study area was usually ≥3, although fewer visits were allowed in rare cases in which females either had no brood patch during the nesting season, or were observed for ≥30 min during the period when they should have been in the late incubation or early brooding stage, and showed no sign of nesting.

In most study areas, there were some Spotted Owl territories that were known from historical surveys before the studies began, but there were also many areas that had never been surveyed and where occupancy by Spotted Owls had never been reported. Because it took several years for surveyors to become familiar with their study areas and to locate and band the territorial owls within their study areas, we truncated the data to exclude the first 1 to 5 years of data on individual study areas. Truncation reduced the number of years in the sampling period, but eliminated some problems with small sample size and incomplete surveys in the early years on each study area. Once surveys began and a sample of owls was banded, new owls entered the study population when they were first detected and banded within the study area.

If owls were located on any of the visits to a given survey area, we followed a standard protocol to document the number of young fledged (NYF) by each female (Lint et al. 1999). The Lint et al. protocol took advantage of the fact that Spotted Owls are relatively unafraid of humans and will readily take live mice from human observers and carry the mice to their nest or fledged young (Lint et al. 1999, Reid et al. 1999). Except in the rare cases mentioned above, our protocol required that owls be located and offered ≥3 mice on two or more occasions each year to document their nesting status and the number of young that left the nest or "fledged" (NYF). If owls ate or cached all the mice offered, and no juvenile owls were detected, then pairs were considered to be non-nesting or failed nesters and were assigned a score of "0" for NYF. For owls that produced ≥1 young, the NYF was coded as the maximum number of young observed on at least two visits after the juveniles left the nest tree. The protocol included some exceptions that we adopted to reduce bias in fecundity estimates. For example, females were given a "0" for NYF if they (1) appeared to be non-nesting based on one or more visits during the spring and then could not be relocated on multiple return visits or (2) were determined to be nesting but could not be relocated on repeated visits to the area. We included these exceptions in our fecundity estimates because females that did not nest and females that nested but failed to produce young sometimes disappeared before the full protocol could be met, and excluding these birds would have caused a positive bias in fecundity estimates. Reproductive data from owls that did not meet the above protocols were recorded as "unknown" and excluded from our analyses.

ANALYTICAL METHODS

Development of Covariates

Barred Owl Covariate

We hypothesized that the presence of Barred Owls near areas occupied by Spotted Owls could have a negative effect on detectability, fecundity, survival, recruitment, or rate of population change of Spotted Owls within our study areas (Kelly et al. 2003, Olson et al. 2005). We did not specifically target Barred Owls in our surveys, but frequently heard or saw Barred Owls while conducting surveys for Spotted Owls, and we recorded the dates and locations of all such detections. The Barred Owl covariate that we used to evaluate our hypotheses was the annual proportion of Spotted Owl territories in each study area that had Barred Owls detected within a 1-km radius of the annual activity centers that were currently or historically occupied by the Spotted Owls on each territory. Consequently, the Barred Owl covariate was a random effect, time (year)-specific variable in analyses of individual study areas that was applied at the scale of the study area or owl population, not individual territories. In meta-analyses of survival and population change (λ), the Barred Owl covariate was a random effects variable that was applied at the meta-population level, but with data that were specific to each study area.

To develop the Barred Owl covariate, we identified an annual "activity center" for each Spotted Owl territory based on the most biologically significant records of the year, ranked in order of declining importance as follows: (1) active nest, (2) fledged young, (3) primary roost, (4) diurnal location, (5) nocturnal response to playbacks, or (6) most recent activity center if no Spotted Owls were located. The territory-specific frame of reference for this analysis was the cumulative area encompassed by 1-km-radius circles around all of the annual activity centers at each Spotted Owl territory. If there was only a single activity center within a territory in all years of the study, then the frame of reference was a single 1-km circle. If there were multiple activity centers used in different years in the same territory, then the frame of reference was the cumulative area encompassed by 1-km-radius circles around all of the annual activity centers within the territory. If Barred Owls were detected anywhere within the cumulative frame of reference in a given year, then that territory was considered to be occupied by Barred Owls in that year, and the annual study area covariate was the proportion of Spotted Owl territories occupied by Barred Owls (Appendix B). We felt that this approach was the best indicator of whether there was likely to be a Barred Owl effect on the Spotted Owls that occupied each territory. Preliminary results indicated that the relative abundance of Barred Owls varied considerably among years and study areas, and that the appearance of Barred Owls in any appreciable numbers on the study areas occurred in Washington in the mid-1980s, Oregon in the early 1990s, and California in the mid-1990s. Consequently, we predicted that any associations between demographic rates of Spotted Owl and Barred Owl detections would be variable among study areas.

Habitat Covariates

Another objective of our analysis was to determine if fecundity, survival, or recruitment were related to the annual percent cover of suitable owl habitat within or adjacent to individual study areas. The frame of reference for habitat covariates in the analysis of fecundity, apparent survival, and recruitment was the percent cover of suitable habitat within each study area. For this estimate, we used a 2.4-km radius around all historical owl activity centers to define each study area (Fig. 2, Appendix C). The acronym used for this environmental covariate was "HAB1." Choice of the 2.4-km radius as the criteria for defining study area boundaries was based on an approximation of the annual area used by resident pairs of Northern Spotted Owls (Forsman et al. 1984, 2005; Carey et al. 1992; Hamer et al. 2007). Although annual home ranges of Spotted Owls vary widely among geographic regions, we opted to simplify

the analysis by using a constant radius to define all study areas.

Our definition of suitable habitat was based on Davis and Lint (2005), who created a base map of suitable Spotted Owl habitat for Washington and Oregon based on multiple covariates, including tree diameter, stand structure, canopy cover, and elevation. Accuracy assessments of these maps were conducted at both the physiographic province and territory scale. At the province scale, maps correlated well with locations of known owl territories, with Spearman rank coefficients ranging from $r_s = 0.83$ to 0.99 ($P < 0.001$; Appendix E in Lint 2005). At the territory scale, 19 sets of independent data from radio-marked Spotted Owls in Oregon indicated that average Spearman rank correlations between suitable habitat and locations of owls were 0.99 in the Coast Ranges, 0.93 in the western Cascades, and 0.94 in the southern Oregon Cascades (Appendix F in Lint 2005). Although there were exceptions, the majority of forests that fit the Davis and Lint (2005) definition of suitable habitat were characterized by large overstory conifers (dbh > 50 cm) and high (>70%) canopy cover (e.g., see Table 3-3 in Davis and Lint 2005:41). The Davis and Lint definition of "suitable owl habitat" does not perfectly define suitable habitat for Northern Spotted Owls throughout their geographic range, but was the best and most current habitat map that was available for our study areas in Oregon and Washington.

Because the base map created by Davis and Lint was based on a single snapshot in time (1996), we used time period-specific stand replacement/disturbance data (Cohen et al. 1998, Healey et al. 2008) to add or subtract habitat in the base map to create a time series of habitat maps for each study area in Oregon and Washington, with four-year time steps in 1984, 1988, 1992, 1996, 2000, and 2002. To create this time series, we assumed that "change" represented loss of habitat, and that the time scale was too short for regrowth of habitat. Therefore, the historical time step maps could be created by "adding back" habitat to the baseline map in years prior to 1996 and subtracting

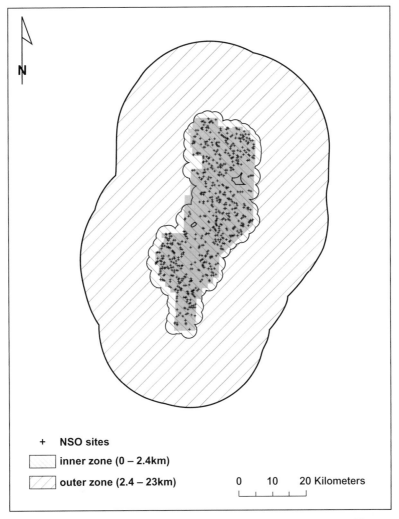

+ NSO sites

inner zone (0 – 2.4km)

outer zone (2.4 – 23km)

0   10   20 Kilometers

Figure 2. Example illustrating frames of reference used to evaluate the proportion of the landscape covered by suitable owl habitat on one of the Northern Spotted Owl demographic study areas (in gray). The small polygon indicates the area within 2.4-km-radius circles around all owl site centers, and the larger polygon indicates the area within 23-km-radius circles around all owl site centers, exclusive of the area of the inner polygon.

habitat from the base map in the years after 1996. To produce annual estimates of suitable habitat, we plotted the estimated percent cover of suitable owl habitat in each time step and then estimated the percent cover of habitat in the years between time steps by assuming a linear trend between the 4-year intervals (Appendix C). Consequently, the habitat covariate was a random effects variable that was time (year)-specific and was applied at the scale of each study area or owl population, comparable to the Barred Owl covariate. For the meta-analyses of survival and λ, the habitat covariate was a random effects

variable that was applied at the meta-population level, with population data that were specific to each study area.

For the habitat covariate in the analysis of λ, we used the same definition of suitable habitat as in the analysis of survival, but developed two covariates based on different spatial scales. One covariate (HAB2) was the same as the HAB1 covariate in the analysis of survival (2.4-km-radius scale), with minor differences due to the fact that we truncated the time-series data to use fewer years in the meta-analysis than the analyses of survival and fecundity on some individual study areas. The second covariate (HAB3) was based on the percent cover of suitable habitat within a 23-km radius of all historical owl activity centers minus the area in HAB2 (Fig. 2, Appendix C). We used the 23-km radius to account for the possible influence of habitat on recruitment from the region immediately surrounding the study area out to a distance that approximated the median natal dispersal distances of Northern Spotted Owls, which were about 19 km for males and 27 km for females (Forsman et al. 2002:15).

After reviewing the habitat map for California, we decided not to develop habitat covariates for study areas from the state map of California because of inconsistencies with the map for Washington and Oregon (Davis and Lint 2005). Two primary problems with the California habitat data were that (1) the California map was based on different remote-sensed data than the combined map for Oregon and Washington (Davis and Lint 2005), and (2) complete evaluation of habitat change in California was not possible because the change detection information for California dated back to only 1994. Therefore, rather than confound our results with maps that were not comparable, we opted to limit our examination of the effects of habitat covariates to Oregon and Washington.

Weather and Climate Covariates

To determine if fecundity, apparent annual survival, or rate of population change were associated with variation in weather and climate, we used climate covariates that were associated with demographic performance of Spotted Owls in previous studies, including mean precipitation and temperature, Palmer Drought Severity Index (PDSI), Southern Oscillation Index (SOI), and Pacific Decadal Oscillation (PDO; Franklin et al. 2000, Seamans et al. 2002, LaHaye et al. 2004, Olson et al. 2004, Dugger et al. 2005, Glenn 2009). These climate variables included measures of seasonal and annual weather as well as longer-term measures of climatic conditions.

We obtained mean temperature and precipitation data for each study area from Parameter Elevated Regression on Independent Slope Models (PRISM) maps (Oregon Climate Service 2008). PRISM maps were developed using weather station data and a digital elevation model to generate raster-based digital maps with 4-km^2 resolution of mean monthly temperature (minimum and maximum) and precipitation on each study area (Daly 2006). We combined the monthly maps into seasonal and annual maps that corresponded with important life history stages of the owl, including winter (1 Nov to 28 Feb), early nesting season (1 Mar to 30 Apr), late nesting season (1 May to 30 Jun), and annual periods (1 Jul to 30 Jun). Temperature and precipitation values for each study area and time period were obtained by computing the average values of raster cells for each seasonal or annual map that fell within study area boundaries.

We used the Palmer Drought Severity Index (PDSI) as an index of primary productivity that has the potential to influence abundance of Spotted Owl prey (NOAA 2008a). The PDSI is the deviation of moisture conditions from normal (30-yr mean = 1970 to 2000), standardized so comparisons can be made across regions and over time (Alley 1984). Values ranged from -6 (extreme drought) to +6 (extremely wet), with zero representing near-normal conditions. The index was calculated separately for climate regions within each state. Most study areas fell within one climate region. For study areas that included multiple climate regions, we used a weighted average of PDSI values based on the proportion of the study area that fell within each climate region.

We used monthly values of the Southern Oscillation/el Niño Index (SOI; NOAA 2008b) and the Pacific Decadal Oscillation (PDO; University of Washington 2008) to assess region-wide climate patterns. We averaged monthly values to obtain annual (Jul 1 to Jun 30) measures of SOI and PDO. Consequently, all of the weather and climate covariates were random effects variables that were time-specific and were applied at the scale of owl populations in the analyses of individual study areas. For the meta-analyses of fecundity, survival, and λ, the weather covariates were random effects variables that were applied at the meta-population level, but with data that were specific to each study area.

Land Ownership, Ecoregion, and Latitude Covariates

To evaluate whether vital rates or rates of population change differed depending on land ownership, ecoregion, or latitude, we developed covariates for land ownership (OWN), ecoregion (ECO), and latitude (LAT). The ownership covariate was a categorical variable that divided study areas into three categories depending on whether they were privately owned, federally owned, or included an approximately equal mix of private and federal ownership (Appendix A). The ecoregion covariate categorized each study area into one of five ecoregions that incorporated geographic location (state) and the major forest type in each study area (Appendix A). Latitude was a continuous variable measured at the center of each study area. In the meta-analyses of fecundity, survival, and λ, all of these covariates were fixed effects variables that were applied at the scale of meta-populations.

Reproduction Covariate

To determine if there was evidence for a cost of reproduction on adult survival in the following year, we used the mean number of young fledged per female as a year- and study area–specific covariate in analyses of apparent survival. We also used the mean NYF covariate in recapture models to investigate the effect of reproduction on detection probabilities in the current year. The mean NYF covariate was time (year)-specific and used as a random effects variable at the scale of populations, comparable to the way we used the Barred Owl and habitat covariates. In the meta-analysis of survival, the NYF covariate was applied at the scale of meta-populations.

Fecundity

Individual Study Areas

We conducted all analyses of reproduction based on the annual number of young produced per territorial female (NYF), but to be consistent with previous reports (Forsman et al. 1996a, Franklin et al. 2004, Anthony et al. 2006), we present the data as "fecundity," where fecundity is the average annual number of female young produced per female owl. We estimated fecundity as NYF/2, based on genetic evidence from blood samples of juveniles that the sex ratio of Spotted Owls is 1:1 at hatching (Fleming et al. 1996). We assumed that the owls in our samples were representative of the population of territorial birds and that sampling was not biased toward birds that reproduced. We think these assumptions were reasonable because Spotted Owls typically occupy the same areas year after year and are reasonably easy to find even in years when they do not breed (Franklin et al. 1996, Reid et al. 1999).

For the analysis of individual study areas, we used PROC MIXED in SAS (SAS Institute, Inc. 2008) to fit a suite of *a priori* models for each study area that included: (1) the effects of age (A), (2) general time variation (t), (3) linear (T) or quadratic (TT) time trends, (4) the proportion of Spotted Owl territories where Barred Owls were detected each year on each study area (BO; see Appendix B), and (5) an even–odd year effect (EO). In addition, we included a simple autoregressive time effect model and the climate and habitat covariates described above (see also Appendix C). The autoregressive time effect model [AR(1)] fits a time trend but allows residuals to be non-independent where $Y_t = \beta_0 + \beta_1 t + \varepsilon_t$ and the correlation of ε_t and $\varepsilon_{t+k} = \rho^k$. Model ranking and selection of best models were based on minimum AIC_c (Burnham and Anderson 2002).

Plots of the annual variance-to-mean ratio for all study areas confirmed that the variance of NYF was nearly proportional to the mean of NYF, with some evidence of smaller variances at higher levels of reproduction. This pattern was consistent with a truncated Poisson distribution

(Evans et al. 1993) because Spotted Owls seldom raise more than two young. However, despite the integer nature of the underlying data (0, 1, 2, and rarely 3 young), the average annual number of young fledged per age class in each study area in each year was not distributed as Poisson (Franklin et al. 1999, 2000; Anthony et al. 2006). For this reason, we did not use a Poisson regression because it is not robust to departures from a Poisson distribution (White and Bennetts 1996). Instead, we used regression models based on the normal distribution, which are less biased when distributions depart from normal. Sample sizes were also sufficiently large to justify the assumption of a normal distribution for each average as long as an allowance was made for the dependence of the variation on the mean (see below; Anthony et al. 2006). The process of averaging NYF also clarified the definition of the sampling unit for this analysis, as the appropriate sample unit was not the individual owl, but the study area–age class combination, which responds to yearly effects that influence the entire study area. Autocorrelation issues in NYF over time for a particular territory were also avoided by treating study areas as the sampling unit. For all these reasons, we used the normal regression model on the annual averages for the analysis of NYF in each age class.

We also reduced the effect of the variance-to-mean relation by fitting models to the annual mean NYF by age class. Annual means for each study area were modeled as

PROC MIXED; MODEL MEAN_NYF = fixed effects.

Thus, residual variation was a combination of year-to-year variation in the actual mean and variation estimated around the actual mean and is approximately equal to

$$\text{var(residual)} = \text{var(yr effects)} + \text{var(NYF)}/n,$$

where n = number of territorial females checked in a particular year. Our approach was justified for several reasons. First, we performed a variance components analysis in which we looked at the individual fecundity records of adult females and estimated the resulting variance components after adjusting for the obvious even–odd year effects. Because Spotted Owls are highly territorial and long-lived, it is difficult to distinguish between spatial and individual effects, and such effects are termed "spatial" components in this report. Our variance components analysis showed that when comparing components of variance, spatial variance among territories tended to be small relative to temporal variance among years and other residual effects (see Results). Therefore, we concluded that ignoring spatial variance within study areas would not bias the results, which negated the need to include owl territory as a random effect. Second, we were able to support the key assumption that the var(residual) was relatively constant because (1) var(NYF)/n was small relative to var(yr effects); (2) the total number of females sampled was roughly constant over time for each study area so that var(NYF)/n was roughly constant; and (3) relatively few (<10%) territorial subadults were encountered, such that var(NYF)/n was also about constant even though var(NYF) may decline with increasing age class. The assumptions were verified by Levene's test for homogeneity of variances (Ramsey and Schafer 2002). Third, we assumed that residual effects were approximately normally distributed because, based on the central limit theorem, the average of the measurements will have an approximate normal distribution with large sample sizes even if the individual measurements are discrete. Finally, covariates included in the analysis of each study area (such as BO) were more easily modeled at the study area (population) level with the above approach.

The best model was not consistent among study areas, so we used a nonparametric approach to estimate mean NYF. First, we computed mean NYF for each year and age class. Then we averaged the means across years within each age class. The estimated standard error was computed as the standard error of the average of the averages among years. This method for estimating NYF gave equal weight to all years, regardless of the number of birds actually sampled in a year, and did not force a model for changes over time. It treated years as random effects with year effects being large relative to within-year-sampling variation. Estimates weighted by sample sizes in each year were not substantially different.

Meta-analysis of Fecundity

In the meta-analysis of fecundity, we restricted the analysis to adult females only because samples of 1- and 2-yr-old owls were small (<10%) in most data sets. In this analysis, we developed a set of a priori models similar to those developed for individual study areas, but in addition to the effects included in the models for individual study areas, we also investigated the effects of latitude (LAT), ecoregion (ECO), and land ownership (OWN; Appendix A) as fixed random variables. We did not have habitat covariates for study areas in California, so we conducted two separate meta-analyses of fecundity. The first analysis included all study areas without any habitat covariates, and the second included study areas from Washington and Oregon only, with habitat covariates included in the a priori models.

We used mixed models to perform meta-analyses on mean NYF per year for the same reasons specified above for the study area analysis. An ecoregion by year (ECO*yr) treatment combination was defined for each study area with owls within study areas as units of measure. Thus, sampling units were study areas within ECO*yr, which we treated as a random effect in the mixed models. Because ownership, latitude, and ecoregion apply at the study-area level rather than at the individual level, we conducted model selection based on average NYF by study area and year. Model rankings and selection of best models were based on minimum AIC_c or $QAIC_c$ (Burnham and Anderson 2002).

Apparent Survival

Individual Study Areas

We used capture–recapture (re-sighting) data to estimate recapture probabilities (p) and annual apparent survival probabilities (φ) of territorial owls. Recapture probabilities were defined as the probability that an owl alive in year $t + 1$ is recaptured, given that it is alive and on the study area at the beginning of year t. Apparent survival was defined as the probability that an owl survives and stays on the study area from time t to $t + 1$, given that it is alive at the beginning of year t. Our general approach for estimating apparent survival was to first develop a set of a priori models for analysis based on biological hypotheses that were discussed and agreed upon by all participants at the workshop. The a priori models were then represented by statistical models in program MARK (White and Burnham 1999). Then we evaluated goodness-of-fit and estimated an overdispersion parameter ($\hat{c}$) for each data set, and estimated recapture probabilities and apparent survival for each capture–recapture data set with the a priori models in program MARK. If needed, we adjusted the covariance matrices and AIC_c values with $\hat{c}$ to inflate variances of parameter estimates and obtain $QAIC_c$ values for model selection. Then, we selected the most parsimonious model for inference based on the $QAIC_c$ model selection procedure (Burnham and Anderson 2002). Additional details on methods of estimation of survival from capture–recapture data from Northern Spotted Owls are provided by Burnham et al. (1994, 1996) and Anthony et al. (2006).

We used Cormack–Jolly–Seber open population models (Cormack 1964, Jolly 1965, Seber 1965, Burnham et al. 1987, Pollock et al. 1990, Franklin et al. 1996) in program MARK (White and Burnham 1999) to estimate apparent survival of owls for each year. The yearly estimate of apparent survival was roughly from 15 June in year t to 14 June in year $t + 1$, which corresponded with the approximate mid-point of the annual field season in the demographic studies (March or April to August). Owls first banded as subadults or adults were assigned to one of three non-juvenile age classes based on plumage attributes (Forsman 1981, Moen et al. 1991, Franklin et al. 1996). The three age classes were: 1-yr-old subadults (S1), 2-yr-old subadults (S2), and ≥3-yr-old adults (A). We did not estimate juvenile survival rates because estimates of juvenile survival were confounded by permanent emigration caused by natal dispersal (Burnham et al. 1996, Forsman et al. 2002). Although permanent emigration can also cause underestimates of survival for non-juvenile owls, we did not consider this a serious bias because site fidelity of adult Spotted Owls is high and because breeding dispersal is most commonly restricted to short movements between adjacent territories (Forsman et al. 2002).

The goal of the data analysis and model selection process was to find a model from an *a priori* list of models that was best in the sense of Kullback–Leibler information (Burnham and Anderson 2002). Prior to model fitting we used the global model $\varphi(a^*s^*t)$, $p(a^*s^*t)$ to evaluate each data set for goodness-of-fit to the assumptions of the Cormack–Jolly–Seber model using the combined χ^2 values and degrees of freedom for Test 2 and Test 3 from program RELEASE (Lebreton et al. 1992). The global model included estimates of age (a), sex (s) and time (t) effects, plus the interactions among age, sex, and time for both φ and p.

We computed estimates of overdispersion ($\hat{c}$) using the median-$\hat{c}$ procedure in program MARK to determine if there was evidence of overdispersion in the data. In cases where there was evidence of overdispersion, we used estimates of $\hat{c}$ to inflate standard errors and adjust the log-likelihood function for the effects of lack of independence in the data.

For the analysis of survival on the individual study areas, we fit models that included the effects of age, sex, time, time trends (linear, quadratic, autoregressive, change-point, cubic spline), and the annual covariates for reproduction (Appendix D) and Barred Owls (Appendix B). We used cubic spline models to fit flexible trends without specifying their form (Hastie and Tibshirani 1990, Green and Silverman 1994, Venables and Ripley 1999). Spline models provide this flexibility by estimating cubic polynomial trends between a series of four knots (two boundary, two interior) in such a way that the polynomials meet smoothly (i.e., are differentiable) at each knot. Boundary knots were placed at the starting and ending year for each study,

while one interior knot was placed midway between the first year of each study and 2002, and the other interior knot was placed at 2002. Cubic spline models with two interior knots estimated six additional parameters each.

We conducted model selection in three stages. First, we identified the best p structure for the data in each study area by using AIC_c model selection (see below) to choose the best model from among a set of *a priori* recapture models developed during the protocol session. The *a priori* models included 11 models that were the same for all study areas (Appendix E) plus up to three optional "biologist's choice" models that could be included if group leaders wanted to evaluate the effects of unique conditions on their study areas. In this stage, we used the same global structure on φ for all models $[\varphi(g^*s^*t)]$, where "g" indicates study area. Then, in stage two, we applied the best p structure from stage one to 64 *a priori* survival models developed during the protocol session (Appendix F) and used AIC_c model selection to identify the best survival model for each study area. Then, we used the φ structure from the best 2 to 3 models in stage two in combination with the p structure from the best 2 to 3 models in stage one to develop 4 to 9 additional models.

We used maximum likelihood estimation to fit models (Brownie et al. 1978, Burnham et al. 1987) and optimized parameter estimation using program MARK (White and Burnham 1999). We used $QAIC_c$ for model selection (Lebreton et al. 1992, Burnham and Anderson 2002), which is a version of Akaike's Information Criterion (Akaike 1973, 1985; Sakamoto et al. 1986) corrected for small sample bias (Hurvich and Tsai 1989) and overdispersion (Lebreton et al. 1992, Anderson et al. 1994). We computed $QAIC_c$ for each candidate model and selected the model with the lowest $QAIC_c$ value as the best model for inference. We used $\Delta QAIC_c$ values to compare models, where $\Delta QAIC_{ci} = QAIC_{ci} - minQAIC_c$. We used Akaike weights (w_i) (i.e., model probabilities) to address model selection uncertainty and the degree to which ranked models were considered competitive. We also used Akaike weights to compute estimates of time-specific, model-averaged survival rates and their standard errors for each study area (Burnham and Anderson 2002:162). We used model averaging because there were usually several competitive ($\Delta QAIC_c < 2.0$) models for a given data set (Burnham and Anderson 2002).

For each study area, we used the variance components module of program MARK to estimate temporal process variation ($\sigma^2_{temporal}$; White et al. 2001, Burnham and White 2002). Use of variance components allowed us to separate sampling variation (variation attributable to estimating a parameter from a sample) in apparent survival estimates from total process variation. Process variation was decomposed into temporal (parameter variation over time) and spatial (individuals on territories) components.

Meta-analysis of Apparent Survival

The meta-analysis of apparent survival rates was based on capture histories of adult males and females from 11 study areas. Subadults were not included because samples of subadults were small in many study areas, and our objective was to reduce the complexity of the analysis to focus on the main variables of interest, including trends in adult survival and the effects of the Barred Owl, reproduction, weather, and habitat covariates. Apparent survival and recapture probabilities were estimated with the Cormack–Jolly–Seber model using program MARK (White and Burnham 1999). The global model for these analyses was $\varphi(g*s*t)\, p(g*s*t)$, where g was study area, s was sex, and t was time (yr). Goodness-of-fit was assessed with the global model in program RELEASE (Burnham et al. 1987), and the estimate of overdispersion (c) was computed as the average of the $\hat{c}$ estimates from the median-$\hat{c}$ routine for each of the 11 study areas, weighted by the number of owls in each study area analysis. Estimates of overdispersion were used to adjust model selection to $QAIC_c$ and to inflate variance estimates. We initially evaluated eight models of recapture probability [$p(g+t), p(R), p(g+s+t), p(R+s), p([g+t]*s)$,

$p(R*s)$, $p(BO)$, $p(BO+g)$] with a general structure on apparent survival [$\varphi(g*t+s)$], where R indicates the effect of reproduction in the current year and BO indicates the potential effect of Barred Owls. Using the best model structure for p from the initial eight models, we evaluated 15 additional models for apparent survival to determine which combinations of area, sex, time, Barred Owl effects (BO), and reproductive effects (R) minimized the amount of Kullback–Leibler information loss (Appendix G). Sex was then removed from the best model to check for strength of this effect. Then we ran four more models in which the group effect of study area (g) was replaced with the group surrogates OWN, ECO, OWN*ECO, and Latitude (LAT). Next, we added six climate covariates for all study areas and a habitat covariate (HAB1) for study areas in Washington and Oregon. The habitat covariate was added to the base model of $\varphi(g)$ as either an additive (+) or an interactive (*) effect. Comparable habitat data were not available for study areas in California, so the habitat covariate was applied only to study areas in Washington and Oregon. Time variation for California study areas was modeled with an additive time effect (t) instead of habitat. Climate data for the Southern Oscillation Index (SOI), Pacific Decadal Oscillation (PDO), mean amount of precipitation during the early nesting season (ENP), and mean temperature during the early nesting season (ENT) were added to the base model of $\varphi(g)$ as either additive (+) or interactive (*) effects.

After reviewing the results of the above analyses, we concluded that the annual variability in apparent survival was too great for any of the covariates for Barred Owls, reproduction, habitat, or climate to have a measurable effect on the modeling or estimates. Consequently, we used the Method of Moments random effects module (White et al. 2001) in program MARK to do some additional *a posteriori* modeling of apparent survival with the above covariates in order to determine the amount of temporal variability explained by each covariate. We used the general model $\varphi(g*t)\, p(g+s+t)$ in the random effects analysis. To estimate the temporal variation explained by each covariate, a random effects design matrix was used that included the study area effect (g) plus the temporal covariate.

Annual Rate of Population Change (λ)

Individual Study Areas

In the analysis of annual finite rate of population change (λ), we used estimates from the reparameterization of the Jolly–Seber capture–recapture model (λ_{RJS}), which was implemented in program MARK based on the *f*-parameterization of the temporal symmetry models of Pradel (1996; see also Franklin 2001). The rationale for using this parameterization instead of Leslie matrix models was discussed in detail in Franklin et al. (2004) and Anthony et al. (2006). Most importantly, estimates of survival rates for juvenile owls from capture–recapture data are biased low because of extensive emigration from the study areas; losses to natal dispersal lead to negatively biased estimates of λ from Leslie matrix models (Anthony et al. 2006). Since the Pradel (1996) method analyzes capture histories in both a forward and backward manner, it treats mortality, reproduction (recruitment), and movements into and out of the study areas equally, and therefore produces less-biased estimates of λ (see Anthony et al. 2006:11 to 13). The two primary assumptions of the Pradel (1996) method are that study area size is constant and that survey effort is relatively constant in each sampling interval. In other words, owls are not gained or lost because of changes in effort or survey area.

In addition to obtaining annual estimates of λ (λ_t) and trends over time in these estimates, the Pradel model allowed for the decomposition of λ_t into two components, apparent survival (φ) and recruitment (*f*), where:

$$\lambda_t = \varphi_t + f_t$$

Here, φ_t is local apparent survival and reflects both survival of territory holders within study areas and site fidelity of territory holders to study areas. Recruitment (f_t) is the number of new animals in the population at time $t + 1$ per animal in the population at time t and reflects both *in situ* recruitment (individuals born on the study area that become established territory holders) and immigration of recruits from outside the study area. Unfortunately, we were unable to further decompose φ_t and f_t. The complement of adult survival includes losses to death and permanent emigration, whereas recruitment includes immigration of new adults, as well as reproductive rate, survival of young, and ability of young birds to obtain territories. Consequently, the estimates of λ_t accounted for all of the losses and gains in the study area populations during each year. All estimates of λ were truncated at 2006, because parameter estimates for the last two years of study were not estimable. In addition, we removed 1 to 5 of the first years of surveys to eliminate any potential bias in estimates of λ that may have been associated with any artificial population growth associated with initial location and banding of owls that occurred during the first few years of each study (Anthony et al. 2006). Our procedure resulted in truncated data sets for each study area, which satisfied the second assumption of equal sampling effort for the Pradel (1996) method.

Estimates of Realized Population Change

We used the methods of Franklin et al. (2004) to convert estimates of λ_t to estimates of realized population change ($\hat{\Delta}_t$), which is the proportional change in estimated population size relative to population size in the initial year of analysis. We computed annual estimates of realized population change on each study area as

$$\hat{\Delta}_t = \prod_{i=x}^{t-1} \hat{\lambda}_i$$

where x was the year of the first estimated λ_t. To compute 95% confidence intervals for Δ_t, we used a parametric bootstrap algorithm (see Franklin et al. 2004:19) with 1,000 simulations. Under this approach, we used the estimates of annual survival, $\hat{\varphi}_t$, recruitment, $\hat{f}_t$, and recapture probabilities, $\hat{p}_t$, together with an estimate of initial abundance, $\hat{N}_x$, to stochastically generate individual capture histories. Each of the 1,000 generated data sets (sets of capture histories) was then analyzed as data and used to obtain estimates of λ_t and Δ_t, from which empirical confidence intervals were constructed. Specifically, we followed the basic approach of Anthony et al. (2006), where the 95% confidence intervals were based on the ith and jth values of Δ_t arranged in ascending order, where $i = (0.025)(1,000)$ and $j = (0.975)(1,000)$.

Meta-analysis of Annual Rate of Population Change

We used encounter histories from banded territorial owls (subadults and adults) in the meta-analysis of λ from the 11 study areas. In this analysis, we used the most general model [$\varphi(g^*t)$ $p(g^*t)$ $f(g^*t)$] as the basis of the random effects modeling. Our approach permitted inferences about the influence of the various covariates on λ_t, φ_t, and f_t and allowed us to investigate whether φ_t or f_t appeared to covary more closely with λ_t. Modeling results included models in two categories: 45 models in the original *a priori* model set and six additional models developed *a posteriori* after looking at the results of the initial model set (Appendix H). Basically, there was evidence from the ranking of the *a priori* models that two covariates (ecoregions, Barred Owls) were important sources of variation for φ_t and f_t, so we developed six models that included both covariates (see last six models in Appendix H). Thus, our inferences were based on the original members of the model set, but we believe that the two-covariate models that we explored should be considered for future modeling in the next cooperative meta-analysis. As in the analyses of individual study areas, estimates of λ from the meta-analysis were truncated at 2006, because parameters for the last two years of study were not estimable.

Statistical Conventions

We used estimates of regression coefficients (β) and their 95% confidence intervals as evidence of an effect on fecundity, apparent survival, or annual rates of population change by the differ-

ent factors or covariates in models. The sign of the coefficient represented a positive (+) or negative (-) effect of a factor or covariate, and the 95% confidence intervals were used to evaluate the evidence for $\beta < 0.0$ (negative effect) or $\beta > 0.0$ (positive effect). We did not use 95% confidence intervals as strict tests of $\beta = 0.0$, but as measures of precision and general evidence of an effect. For example, if the 95% confidence intervals for a regression coefficient did not overlap 0 and the covariate was included in the best or a competitive model, we concluded that there was "strong evidence" for an effect of that factor or covariate. If the 95% confidence interval overlapped 0 broadly, regardless of the model it occurred in, we concluded that there was "no evidence" for an effect of that factor or covariate. Lastly, if a 95% confidence interval overlapped 0 only slightly, with <10% of the interval above or below 0, we concluded that there was "some evidence" of an effect of that factor or covariate. We attempted to use this approach consistently throughout all of the modeling of fecundity, apparent survival, and annual rate of population change (Anthony et al. 2006).

WORKSHOP PROTOCOLS

Data from the demographic studies of Northern Spotted Owls have been examined in four previous workshops, the results of which have been described in four published reports (Anderson and Burnham 1992, Burnham et al. 1994, Forsman et al. 1996a, Anthony et al. 2006) and one unpublished report (Franklin et al. 1999). Participants in these workshops knew that their data and methods would be subjected to considerable scrutiny, and they developed a transparent and consistent protocol for conducting the analyses (Anderson et al. 1999). We followed the same protocol in our workshop, which was held during 9 to 19 January, 2009. Our first step was to subject the data to a formal error-checking process prior to the workshop to make sure that all data were correctly prepared for analysis and that all participants followed the same field protocols for assessing fecundity and survival of owls. The error-checking process was accomplished by first having the lead biologist on each study area prepare their fecundity files and capture history files in a standardized format for analysis in programs SAS (SAS Institute, Inc. 2008) or MARK (White and Burnham 1999). Then we had each group leader submit the field data forms for a randomly selected sample of 10 records each from their fecundity files and capture history files. If the data were correctly formatted and the field data forms supported the data in the random sample, then the data were approved for analysis. If not, the study area leader was apprised of any problems and asked to review and correct their files before resubmitting another 10 randomly selected records for review. The resampling process was repeated until no errors were found in the random samples from each area. Upon arrival at the workshop, each study area leader signed a form stating that their data had passed the error-checking process and were ready for analysis.

Once at the workshop, the entire group of biologists and analysts met and discussed the plausible hypotheses and developed the protocols and *a priori* models that were used in the analysis (Anderson et al. 1999). The planning part of the workshop involved 2.5 days of discussion, including presentations and discussions regarding the covariates that were available for analysis. Once the protocol session was complete and everyone was in agreement regarding which hypotheses would be used and how they would be modeled, the analysis began, and all participants agreed that, regardless of the outcome, they would not withdraw their data once the analysis started.

RESULTS

Fecundity

Individual Study Areas

Estimates of fecundity (mean number of female young fledged per female per year) were based on 11,450 observations of the number of young produced by territorial females. Female age was an important factor affecting fecundity on all areas (Table 2), with mean fecundity generally lowest for 1-yr-olds (0.070 ± 0.015), intermediate for 2-yr-olds (0.202 ± 0.042), and highest for adults (0.330 ± 0.025; Table 3). Estimates of mean fecundity also varied among study areas (Table 3). The overall composition of the territorial female population across all areas and years was 3.8% 1-yr-olds, 6.1% 2-yr-olds, and 90.1% adults. Mean fecundity of adults and 2-yr-olds was markedly higher on the CLE study area than on all other study areas (Table 3).

In 9 of the 11 study areas, the best model or a competitive model included a biennial pattern of high reproduction in even years and low reproduction in odd years (EO effect; Table 2). However, this even–odd year effect was stronger in some areas than others and appeared to be less prominent in the later years of the study (Fig. 3). In addition, alternative models with other types of time effects on fecundity [T, TT, AR(1)] were competitive with the EO models (Table 2). Thus, no single model adequately explained the annual variation in fecundity across all areas.

Of the 11 study areas, seven (CLE, COA, HJA, TYE, KLA, NWC, GDR) had top models or competitive models that included linear (T) or quadratic (TT) time trends on fecundity (Table 2). The best model that included a linear or quadratic time trend on fecundity is listed for each study area in Table 4, along with the slope coefficients and 95% confidence intervals for each model. Based on 95% confidence intervals for β's that either did not overlap zero or barely overlapped zero (Table 4), we concluded that fecundity was declining in five areas (CLE, KLA, CAS, NWC, GDR), stable in three areas (OLY, TYE, HUP), and increasing in three areas (RAI, COA, HJA). Although the best trend model for CAS was not competitive ($\Delta AIC_c = 6.07$), the 95% confidence interval for the slope coefficient from that model did not include zero, suggesting this was an important, if not the best, effect that we investigated for fecundity on CAS (Table 4). Annual variation in fecundity was high on the Washington study areas compared to study areas in Oregon

TABLE 2

Best model and competing models with $\Delta AIC_c < 2.0$, from the analysis of mean age-specific fecundity for female Northern Spotted Owls on 11 study areas in Washington, Oregon, and California.

Study area	Models[a]	K	$-2\log L$	AIC_c	ΔAIC_c	w_i
Washington						
CLE	A + AR(1)	5	85.1	96.5	0.00	0.24
	A + AR(1) + HAB1	6	84.1	98.1	1.51	0.11
	A + T + AR(1)	6	84.1	98.2	1.69	0.11
RAI	A + EO + ENT	6	33.0	48.5	0.00	0.28
OLY	EO	3	52.1	58.9	0.00	0.22
	A + EO	5	47.7	60.0	1.10	0.13
	EO + HAB1	4	51.3	60.7	1.80	0.09
Oregon						
COA	A + T + AR(1) + HAB1	7	-3.7	13.5	0.00	0.06
	A + EO	5	2.2	13.8	0.30	0.05
	A + EO + HAB1	6	-0.5	13.8	0.30	0.05
	A + EO + ENT	6	-0.5	13.8	0.40	0.05
	A + EO + BO	6	-0.5	13.9	0.40	0.05
	A + AR(1) + HAB1	6	-0.2	14.1	0.60	0.04
	A + EO + T	6	-0.1	14.2	0.70	0.04
	A + T + HAB1	6	-0.1	14.3	0.80	0.04
	A + AR(1)	5	2.9	14.3	1.00	0.04
	A + T + AR(1)	6	0.3	14.6	1.10	0.03
	A + EO + T + HAB1	7	-2.6	14.6	1.10	0.03
	A + EO + SOI + HAB1	7	-2.5	14.7	1.20	0.03
	A + EO + ENP	6	0.7	15.1	1.60	0.03
	A + EO + BO + TT	7	-1.8	15.4	1.90	0.02
	A + TT + EO + AR(1)	8	-4.8	15.4	1.90	0.02
HJA	A + EO + HAB1	6	25.2	39.3	0.00	0.17
	A + EO + BO + HAB1	7	22.6	39.4	0.10	0.16
	A + EO + T + HAB1	7	23.7	40.5	1.20	0.09
	A + EO + LNP + HAB1	7	23.9	40.7	1.40	0.08
TYE	A + AR(1) + HAB1	6	28.2	42.0	0.00	0.19
	A + TT + AR(1) + HAB1	8	22.9	42.0	0.00	0.19
	A + T + AR(1) + HAB1	7	26.1	42.5	0.50	0.15
	A + T + AR(1)	6	28.8	42.6	0.60	0.14
	A + AR(1)	5	32.5	43.7	1.70	0.08
KLA	A + EO + T + HAB1	7	13.0	29.4	0.00	0.07
	A + BO	5	18.8	30.1	0.60	0.05

TABLE 2 (*continued*)

TABLE 2 (CONTINUED)

Study area	Models[a]	K	$-2\log L$	AIC_c	ΔAIC_c	w_i
	A + EO + BO + HAB1	7	13.7	30.1	0.60	0.05
	A + EO + HAB1	6	16.3	30.1	0.60	0.05
	A + EO + BO	6	16.6	30.4	0.90	0.04
	A + TT	6	16.9	30.7	1.30	0.04
	A + BO + HAB1	6	17.0	30.8	1.40	0.04
	A + EO + TT	6	14.4	30.8	1.40	0.04
	A*EO + T + HAB1	9	9.0	31.1	1.70	0.03
	A + EO + BO + T	7	14.9	31.3	1.90	0.03
	A	4	22.5	31.4	1.90	0.02
CAS California	A + EO + ENT + HAB1	7	36.2	52.9	0.00	0.51
NWC	A + T	5	45.4	56.4	0.00	0.18
	A + T + AR(1)	6	43.9	57.3	0.90	0.12
	A*EO + T	8	38.8	57.3	0.93	0.12
	A + TT	6	44.9	58.3	1.94	0.07
	A + EO + T	6	44.9	58.3	1.94	0.07
	A + BO + T	6	44.9	58.3	1.95	0.07
HUP	A + EO + ENT	6	-1.3	13.1	0.00	0.16
	A + PDO	5	2.1	13.8	0.64	0.12
	A + ENT	5	2.3	14.0	0.85	0.10
	A + EO + PDO	6	-0.4	14.0	0.88	0.10
	A + ENP	5	3.2	14.8	1.70	0.07
GDR	A + EO + T	6	-13.1	0.6	0.00	0.28
	A + EO + BO	6	-12.2	1.5	0.91	0.18

[a] Model notation indicates structure for effects of owl age (A), even–odd years (EO), linear time (T), quadratic time (TT), autoregressive time [AR(1)], proportion of territories with Barred Owl detections (BO), percent cover of suitable owl habitat within 2.4 km of owl activity centers (HAB1), early nesting season precipitation (ENP), late nesting season precipitation (LNP), early nesting season temperature (ENT), and Pacific Decadal Oscillation (PDO). Habitat information was not available for California, so we did not fit models with habitat covariates for study areas in California.

and California, which may have made it more difficult to detect trends in Washington (Fig. 3). For example, there were a few years with zero reproduction on the RAI and OLY study areas in Washington, whereas years with no reproduction were rare on study areas in Oregon and were never observed in any of the California study areas (Fig. 3).

Models that included the Barred Owl covariate were part of the top model or competitive models for five study areas (COA, HJA, KLA, NWC, GDR; Table 2). Confidence intervals for the slope coefficients of the Barred Owl effect from the best linear or quadratic time-trend model that included the BO covariate indicated a negative relationship between Barred Owls and fecundity on four study areas (COA, KLA, CAS, GDR) and a positive relationship between Barred Owls and fecundity on one study area (HJA; Table 5). On the other six areas (CLE,

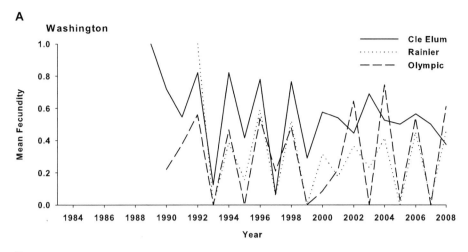

A

Washington

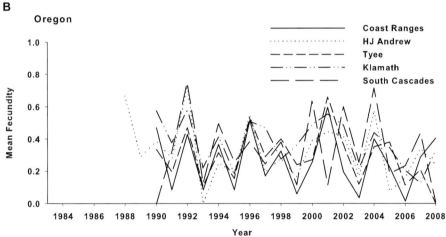

B

Oregon

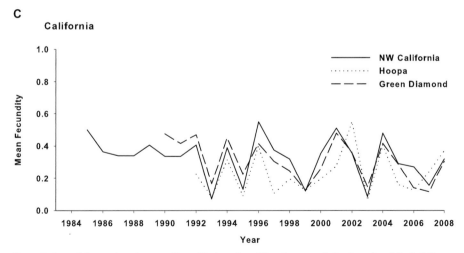

C

California

Figure 3. Annual fluctuations in mean fecundity (number of female young fledged per female) of adult Northern Spotted Owls in three study areas in Washington (A), five study areas in Oregon (B), and three study areas in California (C).

TABLE 3

Estimates of mean fecundity (number of female young produced per female) of Northern Spotted Owls on 11 study areas in Washington, Oregon, and California, subdivided by age class.

Study area	Years	S1			S2			Adults		
		n^a	$\bar{x}$	SE	n^a	$\bar{x}$	SE	n^a	$\bar{x}$	SE
Washington										
CLE	1989–2008	27	0.115	0.083	36	0.517	0.109	499	0.553	0.052
RAI	1992–2008	6	0.100	0.100	11	0.111	0.111	269	0.302	0.065
OLY	1990–2008	8	0.150	0.100	12	0.361	0.162	711	0.300	0.060
Oregon										
COA	1990–2008	25	0.000	0.000	53	0.094	0.039	1,460	0.263	0.040
HJA	1988–2008	15	0.083	0.083	48	0.110	0.043	1,184	0.323	0.041
TYE	1990–2008	67	0.018	0.013	87	0.218	0.065	946	0.305	0.034
KLA	1990–2008	90	0.056	0.024	133	0.289	0.045	1,137	0.377	0.033
CAS	1991–2008	37	0.060	0.038	68	0.210	0.064	1,176	0.347	0.052
California										
NWC	1985–2008	71	0.088	0.054	94	0.152	0.038	1,108	0.324	0.027
HUP	1992–2008	17	0.000	0.000	25	0.077	0.052	377	0.230	0.033
GDR	1990–2008	69	0.095	0.034	126	0.080	0.024	1,458	0.305	0.030
Averages		11	0.070	0.015	11	0.202	0.042	11	0.330	0.025

[a] Sample size indicates the number of cases in which we sampled owls in each age class. This is not a sample that was used to calculate means and standard errors. Those estimates were based on the number of years in the survey period. Estimates were determined using a nonparametric approach. Total number of samples by age class was: S1 = 432, S2 = 693, Adult = 10,325.

RAI, OLY, TYE, NWC, HUP), the 95% confidence intervals on the slope coefficients of the Barred Owl effect broadly overlapped zero, indicating little evidence of an effect of Barred Owls on fecundity (Table 5). In all study areas, the proportion of Spotted Owl territories with Barred Owl detections was increasing with time, but variable among study areas (Appendix B). As a result, temporal trends in fecundity and the Barred Owl covariate were negatively correlated and not easily separated. On some study areas, the temporal effect on fecundity may have been stronger, and this may explain, in part, the lack of effects of Barred Owls on fecundity in some areas. As a result, there was general uncertainty in selection of models with time trends versus Barred Owl effects for most study areas (Table 2).

The habitat covariate (HAB1) was in the top model for all study areas in Oregon, and in competitive models for two of the three study areas in Washington (Table 2). In Oregon, all 95% confidence intervals for regression coefficients for the habitat covariate excluded zero, and on four of the five areas (COA, HJA, TYE, CAS) the habitat effect was positive as predicted, with increased reproductive success associated with increased amounts of suitable habitat. The exception was the KLA study area, where there was evidence that reproductive success declined with increases in suitable habitat (Table 6). On all three study areas in Washington, 95% confidence intervals for the habitat covariate broadly overlapped zero, indicating that there was little evidence for a habitat effect on fecundity on those areas (Table 6).

TABLE 4
Regression coefficients (β̂) for time trends on the mean annual number of young fledged by adult female Northern Spotted Owls in 11 study areas in Washington, Oregon, and California.

Estimates based on the best model containing linear (T), quadratic (TT), or autoregressive [AR(1)] time trends.

Study area	Best model[a]	ΔAIC_c	β̂	SÊ	95% CI Lower	95% CI Upper
Washington						
CLE	A + T + AR(1)	1.69	-0.005	0.006	-0.017	0.006
RAI	A + EO + BO + T	4.49	0.030	0.017	-0.005	0.065
OLY	A + EO + T	3.89	0.004	0.008	-0.014	0.021
Oregon						
COA	A + AR(1) + T + HAB1	0.00	0.070	0.035	-0.001	0.142
HJA	A + EO + T + HAB1	1.22	0.010	0.008	-0.006	0.027
TYE	A + TT + AR(1) + HAB1[b]	0.00	0.106	0.046	0.014	0.197
			0.002	0.001	-0.000	0.004
KLA	A + EO + T + HAB1	0.00	-0.024	0.008	-0.039	-0.008
CAS	A + EO + T	2.34	-0.015	0.005	-0.026	-0.004
California						
NWC	A + T	0.00	-0.009	0.003	-0.015	-0.003
HUP	A + T	4.40	0.005	0.004	-0.004	0.013
GDR	A + EO + T	0.00	-0.007	0.003	-0.012	0.002

[a] Model notation indicates structure for effects of owl age (A), even–odd years (EO), linear time (T), quadratic time (TT), autoregressive time [AR(1)], proportion of territories with Barred Owl detections (BO), percent cover of suitable owl habitat within 2.4 km of owl activity centers (HAB1), early nesting season precipitation (ENP), early nesting season temperature (ENT), and Pacific Decadal Oscillation (PDO). Habitat information was not available for California, so we did not fit models with habitat covariates for study areas in California.

[b] The first estimate is the linear term, and the second is the quadratic term.

Weather or climate covariates occurred in competitive models for RAI, COA, HJA, CAS, and HUP (Table 2), but the best covariate and the direction of the effect varied among areas (Table 7). In particular, the effect of temperature during the early nesting season (ENT) occurred in the top model or a competitive model for four study areas (RAI, COA, CAS, HUP; Table 2). In three of those areas (RAI, COA, CAS), fecundity was positively associated with ENT, as predicted, but the confidence intervals on the slope coefficient for COA included zero (Table 7). In contrast, fecundity was negatively associated with ENT on the HUP study area, which was contrary to what we predicted (Table 7). ENT was also the best climate covari-

ate for GDR, but the model containing ENT was not competitive, and 95% confidence limits on the slope coefficients for the ENT effect included zero (Table 7).

Precipitation during the early nesting season (ENP) occurred in a competitive model for one study area (COA) and was the best weather/climate covariate for CLE and NWC as well (Table 7). The 95% confidence intervals on the slope coefficients for ENP excluded, or just barely included, zero for all three of these study areas, and the association was negative, as predicted (Table 7). There was weak evidence for a negative effect of precipitation on fecundity during the late nesting season (LNP) on the HJA study area, but the 95% confidence interval for

TABLE 5

Regression coefficients (β̂) for the effect of Barred Owls on the mean annual number of young fledged by adult female Northern Spotted Owls in 11 study areas in Washington, Oregon, and California.

Estimates are from the best model that included the Barred Owl (BO) covariate.

Study area	Best model[a]	ΔAIC_c	$\hat{\beta}$	$\hat{SE}$	95% CI Lower	95% CI Upper
Washington						
CLE	A + TT + BO + AR(1)	5.25	0.584	0.983	-1.397	2.566
RAI	A + EO + BO	4.11	-0.505	0.462	-1.455	0.446
OLY	A + EO + BO	4.05	0.045	0.315	-0.601	0.691
Oregon						
COA	A + EO + BO	0.37	-0.137	0.083	-0.305	0.031
HJA	A + EO + BO + HAB1	0.12	0.289	0.176	-0.065	0.643
TYE	A + TT + BO + AR(1) + HAB1	2.34	-0.513	0.726	-1.972	0.946
KLA	A + BO	0.61	-0.459	0.234	-0.928	0.010
CAS	A + EO + BO	7.40	-0.972	0.387	-1.752	-0.193
California						
NWC	A + BO + T	1.95	0.554	0.806	-1.057	2.165
HUP	A + BO	4.88	0.197	0.230	-0.269	0.662
GDR	A + EO + BO	0.91	-0.494	0.203	-0.902	-0.087

[a] Model notation indicates structure for effects of owl age (A), even–odd years (EO), linear time (T), quadratic time (TT), autoregressive time [AR(1)], proportion of territories with Barred Owl detections (BO), percent cover of suitable owl habitat within 2.4 km of owl activity centers (HAB1). Habitat information was not available for California, so we did not fit models with habitat covariates for study areas in California.

the beta coefficient overlapped zero (Table 7). The Southern Oscillation Index (SOI) was the best weather/climate covariate for OLY, but the model that included SOI was not competitive with the best model, and the 95% confidence interval on the slope coefficient overlapped zero (Table 7). The best weather/climate covariate for TYE indicated a negative effect of late nesting season temperature (LNT) on fecundity (Table 7). While this model was not competitive with the best model, the 95% confidence limits on the slope coefficient for the effect of LNT excluded zero, suggesting that temperature during the late nesting season was an important effect and possibly the best predictor of fecundity for TYE.

Estimation of spatial (site-to-site), temporal (year-to-year), and residual variance on the territory-

specific data from the best models indicated that the proportion of variance in number of young fledged attributable to territories and/or individual owls (spatial) was generally <6% (Table 8). The proportion of variance attributable to fluctuations over time was usually in the range of 10 to 20%, while the proportion of unexplained (residual) variation was generally >80%. As a consequence, the explainable variation in fecundity by time and territory was overwhelmed by unexplained, residual variation.

Meta-analysis of Fecundity

The meta-analysis of fecundity for all study areas with no habitat covariates included produced three competitive models (ECO+t, LAT+t, ECO+t+BO), which accounted for 42%,

TABLE 6

Regression coefficients ($\hat{\beta}$) from the best model containing the effect of habitat on the mean annual number of young fledged per adult female Northern Spotted Owl in eight study areas in Washington and Oregon.

Study area	Best model[a]	ΔAIC_c	$\hat{\beta}$	$\hat{SE}$	95% CI Lower	95% CI Upper
Washington						
CLE	A + AR(1) + HAB1	1.5	1.236	1.129	-1.248	3.720
RAI	A + EO + ENT + HAB1	3.2	-1.465	3.832	-9.356	6.426
OLY	EO + HAB1	1.8	-9.253	10.305	-30.300	11.792
Oregon						
COA	A + T + AR(1) + HAB1	0.0	15.672	7.346	0.792	30.552
HJA	A + EO + HAB1	0.0	11.313	2.650	5.787	16.475
TYE	A + AR(1) + HAB1	0.0	0.909	0.432	0.031	1.788
KLA	A + EO + T + HAB1	0.0	8.737	3.415	-15.600	-1.871
CAS	A + EO + ENT + HAB1	0.0	6.066	2.313	1.405	10.727

[a] Model notation indicates structure for effects of owl age (A), even–odd years (EO), linear time (T), autoregressive time [AR(1)], percent cover of suitable owl habitat within 2.4 km of owl activity centers (HAB1), early nesting season temperature (ENT), and forest habitat within 2.4 km radius of owl territory (HAB1). Habitat information was not available for California, so we did not fit models with habitat covariates for study areas in California.

34%, and 19% of the model weights, respectively (Table 9). These three models suggested that fecundity varied by time and was parallel across ecoregions or latitudinal gradients (Fig. 4), with some weak evidence for an additional Barred Owl effect. The estimate of the regression coefficient for the best model with the BO effect was negative, suggesting fecundity decreased as the proportion of territories where Barred Owls were detected increased. However, the 95% confidence interval for the beta coefficient for the BO effect overlapped zero ($\hat{\beta}$ = -0.12, SE = 0.10, 95% CI = -0.31 to 0.07). A linear time trend (T) in fecundity was not supported by the meta-analysis because of the high variation in fecundity over time and the breakdown of the even–odd year effect after about 1999 (Fig. 4). The ΔAIC_c estimates for the best models that included ownership (OWN+t) or climate (ECO+ENP) were 8.6 and 79.0, respectively, indicating that ownership and climate covariates explained little of the temporal varia-

bility in fecundity across the range of the Spotted Owl. Average fecundity over all years was similar among ecoregions except for the Washington Mixed-Conifer region, where mean fecundity was 1.7 to 2.0 times greater than in the other ecoregions (Table 10). Fecundity was lowest for the Oregon Coastal Douglas-fir ecoregion.

The meta-analysis of fecundity for Washington and Oregon, which included the habitat covariate, resulted in two competitive models (ECO+t, ECO+t+HAB1) and a third model that was only slightly less competitive (ECO+t+BO; Table 9). These three models accounted for 55%, 21%, and 17% of the model weights, respectively, and were similar to the most competitive models from the meta-analysis of all study areas, except for the competitive model that included the habitat covariate (Table 9). As in the meta-analysis of all areas, there was some evidence for a weak negative effect of Barred Owls on fecundity, although the 95% confidence

TABLE 7

Regression coefficients ($\hat{\beta}$) from the best model containing the effect of a climate or weather covariate on the mean annual number of young fledged by adult female Northern Spotted Owls in 11 study areas in Washington, Oregon, and California.

Study area	Best model[a]	ΔAIC_c	$\hat{\beta}$	$\hat{SE}$	95% CI Lower	95% CI Upper
Washington						
CLE	A + ENP	2.57	-0.015	0.005	-0.025	-0.004
RAI	A + EO + ENT	0.00	0.091	0.038	0.013	0.169
OLY	A + EO + SOI	3.06	-0.061	0.060	-0.183	0.062
Oregon						
COA	A + EO + ENT	0.34	0.030	0.018	-0.007	0.067
HJA	A + EO + LNP + HAB1	1.39	-0.004	0.003	-0.011	0.003
TYE	A + LNT	7.45	-0.053	0.025	-0.103	-0.004
KLA	A + ENP	2.22	-0.002	0.001	-0.004	0.001
CAS	A + EO + ENT + HAB1	0.00	0.071	0.024	0.022	0.120
California						
NWC	A + ENP	5.12	-0.002	0.001	-0.004	0.000
HUP	A + EO + ENT	0.00	-0.060	0.024	-0.109	-0.011
GDR	A + EO + ENT	4.69	0.023	0.017	-0.011	0.056

[a] Model notation indicates structure for effects of owl age (A), even–odd years (EO), percent cover of suitable owl habitat within 2.4 km of owl activity centers (HAB1), early nesting season precipitation (ENP), early nesting season temperature (ENT), late nesting season temperature (LNT), and Southern Oscillation Index (SOI). Habitat information was not available for California, so we did not fit models with habitat covariates for study areas in California.

interval for the beta coefficient for the effect of Barred Owls overlapped zero ($\hat{\beta}$ = -0.104, SE = 0.129, 95% CI = -0.369 to 0.151). There was no evidence for an effect of habitat on fecundity in the meta-analysis ($\hat{\beta}$ = -0.469, SE = 0.453, 95% CI = -1.363 to 0.426). Linear time trends (T) in fecundity had little support, and models that included ownership (OWN+t) or climate (ECO+ENP+HAB1) were not competitive with the top model (ΔAIC_c = 12.9 and 55.1, respectively).

Apparent Survival

Individual Study Areas

To estimate annual apparent survival we used a sample of 5,244 banded owls, including 796 (15.2%) 1-yr-old subadults, 903 (17.2%) 2-yr-old subadults, and 3,545 (67.6%) adults (Table 1).

The total number of recaptures/resightings of banded owls (19,164) was approximately four times the number of initial captures. The overall χ^2 goodness-of-fit for the global model from program RELEASE summed across study areas was 1,543.2 with 972 degrees of freedom (χ^2 = 1.59, $P > 0.10$), indicating good fit of the data to the Cormack–Jolly–Seber open population mark-recapture model (Table 11). The range of χ^2 for the individual study areas was 0.86 to 2.79, with df ranging from 63 to 125 (Table 11), again indicating good fit to the model for most study areas. Examination of the data indicated that the small lack-of-fit to the Cormack–Jolly–Seber open population model was due primarily to temporary emigration, when owls moved off of the study area for one or more years and later returned or were temporarily displaced as a

TABLE 8

Variance components of the mean annual number of young fledged by adult female
Northern Spotted Owls from a mixed-model analysis of year- and territory-specific estimates.

Study area	Spatial[a]		Temporal[b]		Residual		Total Estimate
	Estimate	% Total	Estimate	% Total	Estimate	% Total	
Washington							
CLE	0.054	6	0.144	16	0.691	77	0.898
RAI	0.000	0	0.009	2	0.453	97	0.467
OLY	0.005	1	0.109	21	0.399	77	0.518
Oregon							
COA	0.006	1	0.102	17	0.486	81	0.600
HJA	0.000	0	0.084	12	0.604	86	0.702
TYE	0.014	2	0.075	11	0.587	86	0.683
KLA	0.015	2	0.051	7	0.661	90	0.734
CAS	0.015	2	0.118	16	0.592	80	0.740
California							
NWC	0.007	1	0.043	6	0.647	91	0.711
HUP	0.021	4	0.016	3	0.481	92	0.523
GDR	0.013	2	0.040	6	0.605	91	0.665

[a] Spatial process variance is the random effects estimate of territory variability.
[b] Temporal process variance is the random effects estimate of annual variability.

TABLE 9

Model selection results from meta-analyses of the annual number
of young fledged per adult female Northern Spotted Owl.

Only models with $\Delta AIC_c < 10$ are shown.

Models[a]	K	$-2\log L$	AIC_c	ΔAIC_c	w_i
All study areas					
ECO + t	31	25.3	98.4	0.0	0.42
LAT + t	27	36.3	98.8	0.4	0.34
ECO + t + BO	32	24.1	99.9	1.6	0.19
t	26	44.5	104.1	5.7	0.04
OWN + t	29	42.4	104.6	8.6	0.01
Washington and Oregon study areas only					
ECO + t	26	34.6	97.9	0.0	0.55
ECO + t + HAB1	27	33.6	99.7	1.9	0.21
ECO + t + BO	27	34.0	100.2	2.3	0.17
ECO + t + BO + HAB1	28	33.2	102.3	4.5	0.06

[a] Model notation indicates structure for effects of ecoregion (ECO), general time (t), proportion of territories with Barred Owl detections (BO), ownership (OWN), and percent cover of suitable owl habitat within 2.4 km of owl activity centers (HAB1).

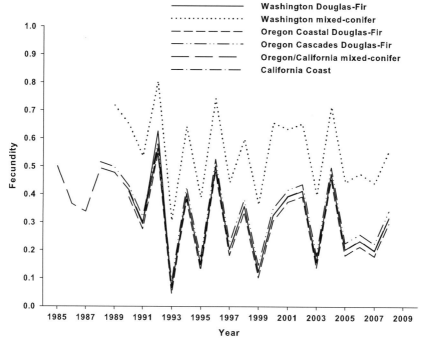

Figure 4. Mean annual fecundity (no. of female young fledged per female) of adult Northern Spotted Owls by ecoregion. Estimates are based on the best model (ECO+t) from a meta-analysis of 11 study areas, where t represents annual time effects and ECO represents the ecoregion effects.

TABLE 10

Estimates of mean annual fecundity (number of female young produced per female) for adult Northern Spotted Owls in six ecoregions.

Ecoregion	$\bar{x}$	SE	95% CI Lower	95% CI Upper
Washington Douglas-fir	0.301	0.043	0.217	0.385
Washington Mixed-conifer	0.553	0.052	0.451	0.655
Oregon Coastal Douglas-fir	0.284	0.026	0.233	0.335
Oregon Cascades Douglas-fir	0.334	0.032	0.271	0.397
Oregon/California Mixed-conifer	0.314	0.019	0.277	0.351
California Coast	0.305	0.030	0.246	0.364

TABLE 11

Estimates of goodness-of-fit and overdispersion ($\hat{c}$) in capture–recapture data for adult Northern Spotted Owls from 11 demographic study areas in Washington, Oregon, and California.

Study area	CJS[a]				λ_{RJS}[a]			
	χ^2	df	χ^2/df	Median-$\hat{c}$	χ^2	df	χ^2/df	Median-$\hat{c}$
Washington								
CLE	72.05	68	1.06	0.99	35.21	51	0.69	1.03
RAI	77.39	72	1.07	1.11	33.73	47	0.72	1.00
OLY	151.50	95	1.59	1.08	156.42	104	1.50	1.04
Oregon								
COA	208.65	97	2.15	1.05	168.87	56	3.02	1.17
HJA	189.38	105	1.80	1.09	167.29	78	2.14	1.09
TYE	90.57	72	1.26	1.04	69.68	64	1.09	1.13
KLA	79.67	92	0.87	1.00	87.48	74	1.18	1.03
CAS	170.94	90	1.90	1.00	142.91	65	2.20	1.06
California								
NWC	76.16	89	0.86	1.00	124.93	81	1.54	1.06
HUP	78.64	63	1.25	0.97	46.06	52	0.89	1.09
GDR	348.25	125	2.79	1.00	139.81	50	2.80	1.00
Totals	1,543.20	972	1.59	1.03[b]	1,366.76	847	1.61	na

[a] CJS indicates data sets used for Cormack–Jolly–Seber estimates of apparent survival. λ_{RJS} indicates data sets used for reparameterized Jolly–Seber estimates of annual finite rate of population growth. Values for χ^2 and df are from TEST 2 and TEST 3 in program RELEASE. Estimates of $\hat{c}$ are from median-$\hat{c}$ routine in program MARK. Estimates of $\hat{c} < 1.0$ were set to 1.00 for analysis.
[b] Weighted average across all study areas.

territorial owl. The overall estimate of overdispersion from the median-$\hat{c}$ routine in program MARK was 1.03, with estimates for individual study areas ranging from 0.97 to 1.11 (Table 11). Overall, results of GOF testing indicated there was little to no overdispersion (i.e., lack of independence) of recaptured owls.

Although there were exceptions, estimates of annual recapture probabilities (p) typically were high, ranging from 0.70 to 0.95 on most study areas. High rates of recaptures/resightings make the Spotted Owl an ideal species for mark–recapture studies. In the analyses of recapture probabilities, factors affecting p in the best models varied among study areas (Table 12). For seven of the 11 areas, there was an effect of sex on p; in all seven cases, p was higher for males. Other effects on p in the top models for one or more areas were a variable time effect (OLY, HJA, CAS areas), negative Barred Owl effect (RAI, COA, KLA areas), and/or a positive reproductive effect (RAI, CLE, TYE areas; Table 12). There was no evidence of time trends on p on any study areas. On two study areas, the "biologist's choice" models were the best models for p. The best p model for one of these areas (NWC) included the additive effects of sex and recapture method; in this case, owls were physically recaptured in 1986 to 1987 and then resighted or recaptured in subsequent years. The other case in which the biologist's choice model was the best p model included an east–west division of the HUP study area based on differences in Spotted Owl density, forest type, and ease of access (Table 12).

TABLE 12

Estimates of model-averaged mean apparent survival ($\hat{\bar{\varphi}}$) for three
age classes of Northern Spotted Owls on 11 study areas in Washington, Oregon, and California.

Study area	Structure on best model[a]	Sex	S1[b]		S2[b]		Adult[b]	
			$\hat{\bar{\varphi}}$	$\hat{\text{SE}}$	$\hat{\bar{\varphi}}$	$\hat{\text{SE}}$	$\hat{\bar{\varphi}}$	$\hat{\text{SE}}$
Washington								
CLE	$\varphi(CP)\ p(R)$	F	0.794	0.051	0.820	0.023	0.819	0.013
		M	0.795	0.051	0.820	0.023	0.819	0.013
RAI	$\varphi[(S1 = S2, A) + BO]\ p(BO + R)$	F	0.541	0.181	0.674	0.156	0.841	0.019
		M	0.546	0.181	0.678	0.157	0.844	0.018
OLY	$\varphi[(S1, S2 = A) + s + T]\ p(s + t)$	F	0.529	0.148	0.786	0.081	0.828	0.016
		M	0.571	0.145	0.814	0.075	0.852	0.014
Oregon								
COA	$\varphi[(S1 + S2 + A) + TT]\ p(BO + s)$	F	0.742	0.072	0.864	0.031	0.859	0.009
		M	0.748	0.071	0.868	0.030	0.863	0.008
HJA	$\varphi[(S1, S2 = A) + t]\ p(s + t)$	F	0.717	0.084	0.830	0.042	0.865	0.010
		M	0.717	0.084	0.830	0.042	0.864	0.010
TYE	$\varphi[(S1, S2 = A) + TT]\ p(R + s)$	F	0.761	0.043	0.864	0.020	0.856	0.008
		M	0.762	0.042	0.865	0.019	0.857	0.008
KLA	$\varphi[(S1, S2 = A) + t]\ p(BO + s)$	F	0.788	0.040	0.858	0.020	0.848	0.008
		M	0.786	0.040	0.857	0.020	0.847	0.008
CAS	$\varphi[(S1, S2 = A) + TT]\ p(t)$	F	0.692	0.069	0.733	0.053	0.851	0.010
		M	0.697	0.069	0.737	0.053	0.853	0.010
California								
NWC	$\varphi[(S1 = S2, A) + T]\ p(Meth + s)$	F	0.774	0.031	0.784	0.031	0.844	0.009
		M	0.776	0.031	0.787	0.031	0.846	0.009
HUP	$\varphi(S1, S2 = A)\ p(EW + Effort)$	F	0.758	0.087	0.838	0.038	0.854	0.014
		M	0.762	0.086	0.840	0.037	0.857	0.013
GDR	$\varphi[(S1, S2 = A) + BO]\ p(s)$	F	0.767	0.044	0.852	0.015	0.853	0.007
		M	0.764	0.045	0.850	0.015	0.851	0.007

[a] Model notation indicates structure for additive (+) or interactive (*) effects of sex (s), time (t), linear time trend (T), quadratic time trend (TT), 2004 change point (CP), reproduction (R), proportion of territories with Barred Owl detections (BO), age class (S1, S2, A), east–west binomial subdivision of study area (EW), survey method (Meth), or differential survey effort in particular years (Effort). An "=" sign means that age classes were combined, and a "," indicates they were modeled separately.

[b] Age classes (S1, S2, A) indicate owls that were 1, 2, or ≥3 years old. Average survival is the arithmetic mean of model-averaged annual survival estimates. Standard errors were calculated using the delta method.

The best model structure for apparent survival (φ) varied among study areas, but several patterns emerged (Table 12). Most notably, apparent survival tended to be higher for adults than for subadults and was similar between the sexes, except on the OLY study area where males had higher survival than females (Table 12). Presence of Barred Owls, variable time (t), or time trends (T or TT) were important effects on apparent survival in one or more study areas. In the best models for each study area (Table 12), the Barred Owl covariate was included in the φ

TABLE 13

Coefficient estimates (β̂) for the best models that included a time trend
on apparent survival of non-juvenile Northern Spotted Owls on 11 study areas
in Washington, Oregon, and California.

Study area	Model trend[a]	ΔQAIC$_c$	β̂	$\widehat{SE}$	95% CI Lower	95% CI Upper
Washington						
CLE	CP (T)[b]	0.00	-0.027	0.021	-0.069	0.015
			-0.182	0.073	-0.324	-0.039
RAI	CP (T)[b]	2.48	-0.143	0.057	-0.254	-0.031
			0.205	0.129	-0.048	0.458
OLY	T	0.00	-0.032	0.016	-0.064	0.000
Oregon						
COA	TT[c]	0.21	0.146	0.046	0.056	0.237
			-0.009	0.002	-0.014	-0.005
HJA	T	0.01	-0.013	0.010	-0.033	0.007
TYE	TT[c]	0.00	0.154	0.048	0.060	0.247
			-0.008	0.002	-0.013	-0.003
KLA	CP[d]	4.38	-0.030	0.025	-0.079	0.020
CAS	TT[c]	0.00	0.169	0.058	0.056	0.282
			-0.009	0.003	-0.015	-0.002
California						
NWC	T	0.00	-0.016	0.008	-0.033	0.000
HUP	CP[d]	1.61	-0.031	0.049	-0.127	0.063
GDR	T	0.54	-0.030	0.009	-0.048	-0.011

[a] T = linear time trend, TT = quadratic time trend, CP = change point starting in 2004.

[b] Models that have a change point beyond which the function changes. The first row estimate is the linear time trend (T) and the second is a change point starting in 2004 (CP).

[c] For quadratic models (TT), the first row indicates the linear term and the second row indicates the quadratic term.

[d] Constant survival from start year to 2004, with negative time trend beginning in 2004.

structure for two study areas (RAI, GDR) and the *p* structure for three study areas (RAI, COA, KLA). The Barred Owl covariate also occurred in competitive models for φ on the OLY and NWC areas (see Effects of Barred Owls on Recapture and Survival below).

Based on the best survival models that included time trends, we concluded that apparent survival was declining on 10 of the 11 study areas (CLE, RAI, OLY, COA, HJA, TYE, CAS, NWC, HUP, GDR), as indicated by 95% confidence intervals on β that either did not overlap

zero or narrowly overlapped zero (Table 13). Declines in apparent survival were most evident in Washington, where all β estimates were negative with 95% confidence intervals that did not overlap zero (Fig. 5A). In addition, the declines in apparent survival on the CLE and RAI study areas were most precipitous during the last five years of the study, as represented by the change-point (CP) time structure in the best models and steeper declines after 2004 (Fig. 5A). Annual estimates of apparent survival for owls on the CLE, RAI, and OLY areas

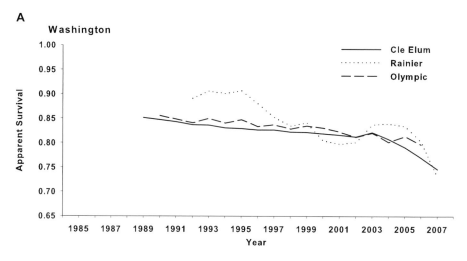

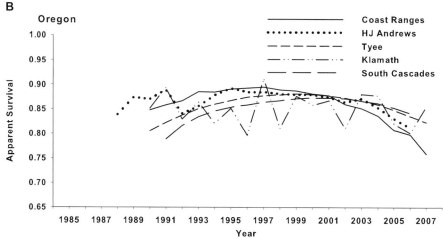

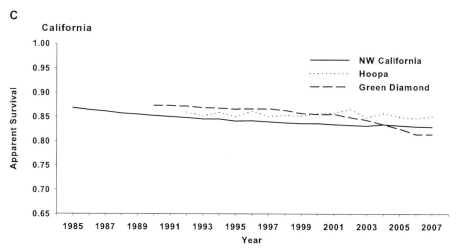

Figure 5. Model averaged estimates of apparent survival of adult female Northern Spotted Owls in three study areas in Washington (A), five study areas in Oregon (B), and three study areas in California (C).

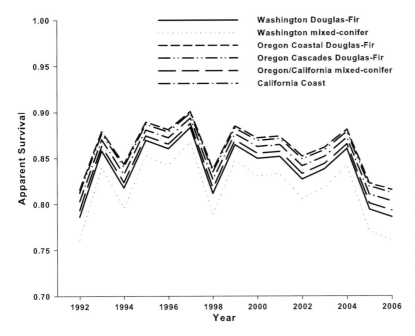

Figure 6. Estimates of apparent annual survival of adult female Northern Spotted Owls in six ecoregions, based on model φ(ECO+t) p(g+t+s) from the meta-analysis of 11 study areas, where ECO represents ecoregion, t represents annual time effects, g represents study area effects, and s represents sex effects.

were <0.80 during the latter years of the study, which were the lowest rates recorded. In Oregon, apparent survival declined on four (COA, HJA, TYE, CAS) of the five study areas, most noticeably during the last five years of study (Fig. 5B). Temporal changes in apparent survival for COA, TYE, and CAS were best described by a quadratic function, whereby survival increased during the early years of the study, then declined during later years. The owl population on the KLA study area was the only one in Oregon that did not have a declining survival rate, as the best model for KLA supported a variable time (t) effect (Table 12). In California, there was strong evidence for linear or change-point declines in apparent survival on all three study areas (NWD, HUP, GDR), as indicated by 95% confidence intervals for β's that either did not overlap zero or only narrowly overlapped zero (Table 13, Fig. 5C).

Meta-analysis of Apparent Survival on All Areas

We used encounter histories from 3,545 adults in the meta-analysis of apparent survival (Table 1). The estimate of goodness-of-fit from program RELEASE indicated good fit of the data

to the Cormack–Jolly–Seber open population model ($\chi^2 = 1740.9$, df = 1,012, $P > 0.10$). The weighted average estimate of median-$\hat{c}$ was 1.031, indicating little overdispersion (i.e., lack of independence) in capture histories. We used this estimate to adjust model selection from AIC_c to $QAIC_c$ and inflate variance estimates accordingly.

The best model from the meta-analysis of apparent survival was the random effects model φ(g*t) p(g+s+t): RE(g+R), which indicated that survival varied among study areas (g) and years (t) and that recapture rates varied among study areas, sexes, and years (Table 14). This model, which had a $QAIC_c$ weight of 0.18, also included the reproduction covariate (R). The effect of reproduction was negative with a 95% confidence interval that barely overlapped zero (Table 15). Several random effects models were competitive, including a second-best model that included the Barred Owl (BO) covariate. The regression coefficient for the BO covariate was negative, with a 95% confidence interval that did not overlap zero (Table 15). For more details on the effects of Barred Owls on apparent survival, see below. Other random effects models with $\Delta QAIC \leq 2$ from the best model were identical in structure

TABLE 14

Model selection criteria for a priori and post hoc models used in the meta-analysis of apparent survival of adult Northern Spotted Owls on 11 demographic study areas in Washington, Oregon, and California, 1985–2008.

Model[a]	K	Q-Deviance[b]	QAIC$_c^c$	ΔQAIC$_c$	w_i
Random effects models					
φ(g*t) p(g + s + t): RE (g+R)	142.9	13,470.07	32,659.14	0.00	0.18
φ(g*t) p(g + s + t): RE (g + BO)	142.1	13,471.89	32,659.33	0.19	0.16
φ(g*t) p(g + s + t): RE (g + BO + PDO)	142.2	13,471.86	32,659.57	0.43	0.14
φ(g*t) p(g + s + t): RE (g + PDO)	143.2	13,470.27	32,659.89	0.75	0.12
φ(g*t) p(g + s + t): RE (g + T)	143.0	13,471.01	32,660.26	1.12	0.10
φ(g*t) p(g + s + t): RE (g + Mean)	143.3	13,470.49	32, 660.45	1.31	0.09
φ(g*t) p(g + s + t): RE (g + ENP)	143.7	13,470.15	32,660.82	1.68	0.08
φ(g*t) p(g + s + t): RE (g + ENT)	143.8	13,470.08	32,660.91	1.77	0.07
φ(g*t) p(g + s + t): RE (g + SOI)	143.9	13,470.04	32.661.06	1.93	0.07
φ(g*t) p(g + s + t): RE (g + HAB1)	205.2	13,460.60	32,776.34	117.02	0.00
Fixed effects models					
φ(ECO + t) p(g + s + t)	62	13,732.87	32,758.61	99.47	0.00
φ(ECO + OWN + t) p(g + s + t)	64	13,730.05	32,759.82	100.68	0.00
φ(g + t) p(g + s + t)	67	13,726.38	32,762.18	103.04	0.00
post hoc φ(g + t + BO) p(g + s + t)	68	13,725.04	32,762.86	103.72	0.00
φ(g + s + t) p(g + s + t)	68	13,725.90	32,763.71	104.57	0.00
post hoc φ(g + t + HAB1) p(g + s + t)	68	13,726.30	32,764.11	104.98	0.00
post hoc φ(g*California + HAB1 + t) p(g + s + t)	61	13,743.14	32,766.87	107.74	0.00
φ(LAT + t) p(g + s + t)	58	13,752.30	32,769.96	110.82	0.00
post hoc φ(t + BO) p(g + s + t)	58	13,752.60	32,770.31	111.17	0.00
φ(OWN + t) p(g + s + t)	59	13,752.80	32,772.54	113.40	0.00
φ(g + BO + s) p(g + s + t)	47	13,830.54	32,826.13	166.99	0.00
φ(ECO + T) p(g + s + t)	41	13,842.81	32,826.35	167.22	0.00
φ(g*R) p(g + s + t)	57	13,812.57	32,828.26	169.12	0.00
φ(ECO*T) p(g + s + t)	46	13,836.97	32,830.55	171.41	0.00
φ(R + s) p(g + s + t)	37	13,856.51	32,832.03	172.89	0.00
φ(g*s*t) p(g*s*t) global	782	12,764.58	33,287.46	628.32	0.00

[a] Codes indicate model structure for additive (+) or interactive (*) effects of ecoregion (ECO), study area (g), sex (s), annual time (t), linear time trend (T), land ownership (OWN), latitude (LAT), proportion of territories with Barred Owl detections (BO), percent cover of suitable owl habitat within 2.4 km of owl activity centers (HAB1), reproduction (R), Pacific Decadal Oscillation (PDO), early nesting precipitation (ENP), early nesting temperature (ENT), or Southern Oscillation Index (SOI).

[b] Q-Deviance is the difference between -2log( )/$\hat{c}$ of the current model and -2log( )/$\hat{c}$ of the saturated model.

[c] $\hat{c}$ values for individual study areas can be found in Table 11.

TABLE 15

Coefficient estimates (β̂) for covariates included in the meta-analysis of apparent
survival of non-juvenile Northern Spotted Owls on 11 study areas in Washington, Oregon, and California.

Covariate	Model[a]	β̂	SE	95% CI Lower	95% CI Upper
Random effects models					
R	φ(g*t) p(g + s + t): RE (g + R)	-0.024	0.013	-0.049	0.001
BO	φ(g*t) p(g + s + t): RE (g + BO)	-0.086	0.037	-0.158	-0.014
PDO	φ(g*t) p(g + s + t): RE (g + PDO)	0.009	0.006	-0.002	0.019
T	φ(g*t) p(g + s + t): RE (g + T)	-0.002	0.001	-0.004	0.000
ENP	φ(g*t) p(g + s + t): RE (g + ENP)	0.000	0.000	-0.001	0.000
ENT	φ(g*t) p(g + s + t): RE (g + ENT)	0.004	0.006	-0.007	0.015
SOI	φ(g*t) p(g + s + t): RE (g + SOI)	-0.002	0.006	-0.014	0.009
HAB1	φ(g*t) p(g + s + t): RE (g + HAB1)	0.339	0.354	-0.352	1.030
Fixed effects models					
Ecoregion[b]	φ(ECO + t) p(g + s + t)				
OR Cascades Douglas-fir		0.162	0.070	0.024	0.300
WA Mixed-conifer		-0.142	0.100	-0.338	0.055
OR-CA Mixed-conifer		0.042	0.070	-0.094	0.179
OR Coast Douglas-fir		0.184	0.071	0.046	0.323
CA Coast		0.103	0.075	-0.044	0.251
Ownership[c]	φ(ECO + OWN + t) p(g + s + t)				
Federal		-0.190	0.115	-0.416	0.036
Mixed		-0.136	0.113	-0.357	0.086
BO	*post hoc* φ(g + t + BO) p(g + s + t)	-0.339	0.293	-0.914	0.237
Habitat	*post hoc* φ(g + t + HAB1) p(g + s + t)	-0.466	1.852	-4.097	3.165
Latitude	φ(LAT + t) p(g + s + t)	-0.009	0.009	-0.026	0.009
Reproduction	φ(R + s) p(g + s + t)	-0.200	0.065	-0.328	-0.072

[a] Codes indicate effects of study area (g), time (t), sex (s), proportion of territories with Barred Owl detections (BO), reproduction (R), Pacific Decadal Oscillation (PDO), linear time trend (T), percent cover of suitable owl habitat within 2.4 km of owl activity centers (HAB1), land ownership (OWN), latitude (LAT), early nesting precipitation (ENP), early nesting temperature (ENT), or Southern Oscillation Index (SOI).

[b] WA Douglas-fir was the reference type.

[c] Non-federal ownership was the reference type.

to the best model, except that the reproduction covariate was replaced by other environmental covariates, including Pacific Decadal Oscillation (PDO), linear time effects (T), mean effects, early nesting season precipitation (ENP), early nesting season temperature (ENT), or Southern Oscillation Index (SOI; Table 14). The random effects models were based on the assumption that the years of our study were a sample of all possible years, whereas the fixed effects models pertained directly to the years sampled. Although none of the fixed effects models were competitive with the best random effects model (Table 14), it is important to describe the results for each analysis because they represent different interpretations of the data (see Methods).

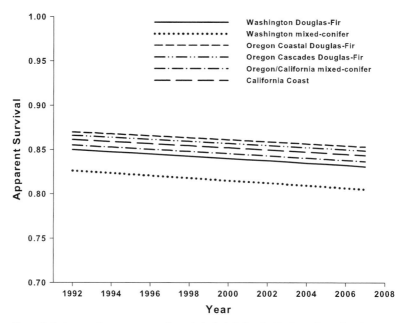

Figure 7. Estimates of apparent annual survival of adult female Northern Spotted Owls in six ecoregions (ECO), based on the linear time-trend model φ(ECO+T) p(g+t+s) from the meta-analysis of 11 study areas. Study area effects are represented by g, annual time effects by t, and sex effects by s.

In the meta-analysis of survival, the best or competing models indicated that there was considerable variation in survival rates among study areas, ecoregions, and years (t), and that the variation in survival among study areas and ecoregions was parallel over time (Fig. 6). Because the general trend in survival suggested a slight decline over the period of study (Fig. 6), we investigated the regression coefficients in the best random effects and fixed effects models that included time trends (T). The best random effects model with a time trend [φ(g*t) p(g+s+t): RE(g+T)] included a negative effect on survival ($\hat{\beta}$ = -0.0016), with a 95% confidence interval that barely overlapped zero (Table 15). The best fixed effects model with a time trend [φ(ECO+T) p(g+s+t)] also provided evidence for an overall decline in apparent survival for all study areas combined (Fig. 7).

Several other covariates were included in competitive models for the meta-analysis of apparent survival. There was no evidence from the random effects models that early nesting season temperature (ENT), Southern Oscillation Index (SOI), or percent cover of suitable owl habitat (HAB1) had an effect on survival because the 95% confidence intervals for these covariates included zero (Table 15). In contrast, there was some evidence that presence of Barred Owls (BO), early nesting season precipitation (ENP), and time trends (T) each had an effect on survival rates in the random effects models (Table 15). From the fixed effects models, there was evidence that survival rates differed among ecoregions, with the Oregon Cascades Douglas-fir, Oregon Coast Douglas-fir, and California Coast regions having higher survival rates than the Oregon/California Mixed-conifer and Washington Mixed-conifer regions (Table 15; Fig. 7). There was no evidence from the fixed effects models that ownership, Barred Owls, habitat, or latitude had an effect on survival, but there was evidence that annual survival was negatively related to the mean number of young produced in the previous breeding season ($\hat{\beta}$ = -0.200, 95% CI = -0.328 to -0.072). Although the evidence suggested that several of the above covariates influenced apparent survival, they explained little (0 to 5.7%,

TABLE 16

Models selected in the meta-analysis of apparent annual survival of Northern
Spotted Owls for eight monitoring areas in Washington, Oregon, and California.

Model[a]	K	Q-Deviance	QAIC$_c$[b]	ΔQAIC$_c$	w_i
Random effects models					
φ(g*t) p(g + s + t): RE (g + R)	152.68	10,811.970	26,028.850	0.000	0.200
φ(g*t) p(g + s + t): RE (g + BO)	152.46	10,812.900	26,029.327	0.473	0.158
φ(g*t) p(g + s + t): RE (g + Mean)	153.00	10,812.210	26,029.745	0.892	0.129
φ(g*t) p(g + s + t): RE (g + PDO)	153.27	10,811.850	26,029.937	1.083	0.117
φ(g*t) p(g + s + t): RE (g + T)	153.23	10,812.130	26,030.132	1.279	0.106
φ(g*t) p(g + s + t): RE (g + ENP)	153.31	10,811.980	26,030.145	1.291	0.105
φ(g*t) p(g + s + t): RE (g + SOI)	153.51	10,811.870	26,030.440	1.586	0.091
φ(g*t) p(g + s + t): RE (g + ENT)	153.51	10,811.880	26,030.461	1.607	0.090
φ(g*t) p(g + s + t): RE (g + HAB1)	157.84	10,809.420	26,036.809	7.956	0.003
Fixed effects models					
φ(ECO + t) p(g + s + t)	58	11,023.270	26,048.455	19.601	0.000
φ(OWN + ECO + t) p(g + s + t)	59	11,022.470	26,049.665	20.811	0.000
φ(g + s + t) p(g + s + t)	62	11,019.080	26,051.603	22.749	0.000
φ(LAT + t) p(g + s + t)	55	11,044.310	26,063.449	34.596	0.000
φ(OWN + t) p(g + s + t)	55	11,044.490	26,063.631	34.778	0.000

[a] Model notation indicates structure for study area (g), time (t), linear time (T), ecoregion (ECO), land ownership (OWN), constant (.), proportion of territories with Barred Owl detections (BO), early nesting season precipitation (ENP), early nesting season temperature (ENT), percent cover of suitable owl habitat within 2.4 km of owl activity centers (HAB1), Southern Oscillation Index (SOI), and Pacific Decadal Oscillation (PDO).

[b] $\hat{c}$ values for individual study areas can be found in Table 11.

individually) of the variation among study areas and years. Thus, there was considerable annual variation in survival estimates (Fig. 6), and no covariate, including Barred Owls, percent cover of suitable habitat, climate, or time trends, explained a major portion of this variation. For example, the Barred Owl covariate and time trend explained only 5.7 and 2.3% of the variability in apparent survival, respectively.

Meta-analysis of Apparent Survival on the Eight NWFP Monitoring Areas

The two best models in the meta-analysis of apparent survival for the eight NWFP study areas were the same as the analysis of all 11 study areas (Table 16). In the top model, the regression coefficient for the effect of reproduction was negative with a 95% confidence interval that barely overlapped zero. In the second best model, the regression coefficient for the effect of Barred Owls was negative with a 95% confidence interval that did not overlap zero. Six other random effects models that were competitive included mean effects, Pacific Decadal Oscillation (PDO), time trend (T), early nesting season precipitation (ENP), Southern Oscillation Index (SOI), or early nesting season temperature (ENT) in place of the BO covariate (Table 16). The rankings of the random effects and fixed effects models were similar between the analyses of all 11 study areas and the eight NWFP monitoring areas, and none of the fixed effects models were competitive with the best random effects models (Tables 14, 16). Because the results were similar regardless of

TABLE 17
Coefficient estimates (β̂) for the best models that included an
effect of reproduction on apparent survival of non-juvenile Northern
Spotted Owls on 11 study areas in Washington, Oregon, and California.

Study area	ΔQAIC$_c$	β̂	ŜE	95% CI Lower	95% CI Upper
Washington					
CLE	2.72	0.466	0.220	0.035	0.897
RAI	2.88	-1.030	0.450	-1.910	-0.014
OLY	0.75	-0.420	0.241	-0.893	0.053
Oregon					
COA	22.96	0.088	0.181	-0.267	0.443
HJA	7.30	-0.165	0.194	-0.546	0.216
TYE	8.33	0.317	0.261	-0.195	0.829
KLA	5.69	0.041	0.214	-0.378	0.461
CAS	7.23	-0.129	0.194	-0.509	0.252
California					
NWC	2.65	0.249	0.234	-0.210	0.708
HUP	0.28	0.573	0.447	-0.304	1.450
GDR	5.16	0.556	0.239	0.088	1.024

whether we examined the eight NWFP study areas or all 11 study areas combined, we emphasize only the results from all 11 areas in the following sections.

Potential Cost of Reproduction on Survival

In the analyses of apparent survival for individual study areas, there was no evidence of a negative effect of reproduction on survival rates in the following year at seven of the 11 study areas (COA, HJA, TYE, KLA, CAS, NWC, HUP, Table 17). Confidence intervals for the regression coefficients for reproduction at those seven areas all overlapped zero (Table 17). For two study areas in Washington (RAI, OLY), there was evidence of a negative effect of reproduction on survival in the following year. At RAI, the regression coefficient for the reproductive effect in the best model was negative with a 95% confidence interval that did not overlap zero. At OLY, the effect of reproduction was part of a competitive model in which the 95% confidence interval on β̂ barely overlapped zero (Table 17). In contrast, there was evidence of a positive effect of reproduction on survival at CLE and GDR, as the regression coefficients for the reproduction covariates were positive, with 95% confidence intervals that did not overlap zero. However, the models for CLE and GDR that included the effect of reproduction were >2 QAICs from the best models, and these latter results were contrary to our original hypothesis.

In the meta-analysis of apparent survival for all 11 study areas, the best random effects model, φ(g*t) p(g+s+t): RE(g+R), included the effect of reproduction. The effect of reproduction was negative (β̂ = -0.024) and the 95% confidence interval barely included zero (-0.049 to 0.001). The best fixed effects models with an effect of reproduction were φ(g*R) p(g+s+t) and φ(R+s) p(g+s+t) (Table 14). Although there was little support for either of these models (ΔQAIC$_c$'s > 168.0 and QAIC$_c$ weights = 0.000),

the regression coefficient for the effect of reproduction in the second model was negative ($\hat{\beta} =$
-0.200) with a 95% confidence interval (-0.328
to -0.072) that did not overlap zero (Table 15).
Based on this outcome, we concluded that there
was evidence for a negative effect of reproduction on survival in the following year in some,
but not all, study areas.

Effects of Barred Owls on Recapture and Survival

The BO covariate was included in the best model
structure for recapture probability in three (RAI,
COA, KLA) of the 11 study areas (Table 12), and
the best models that included a BO effect on
recapture indicated a negative effect in seven
study areas and a positive effect in four areas.
However, the 95% confidence intervals on the
regression coefficients for the BO effect

overlapped zero in seven areas. In the four cases
where the 95% confidence intervals did not overlap zero, two cases indicated a negative effect and
two cases indicated a positive effect.

In the analysis of individual study areas, we
found evidence for a negative effect of Barred
Owl presence on apparent survival of Spotted
Owls on the RAI, COA, HJA, and GDR study
areas (Table 18). There also was some evidence
that presence of Barred Owls had a negative
effect on apparent survival of Spotted Owls on
the OLY and NWC study areas; on those areas
the Barred Owl effect was among the competitive models, but the 95% confidence intervals
for the regression coefficient barely overlapped
zero (Table 18). Inexplicably, there was one
study area (CAS) that had weak evidence for a
positive effect of Barred Owls on survival
(Table 18). The evidence for an effect of Barred

TABLE 18

Estimates of $\Delta QAIC_c$ and parameter estimates ($\hat{\beta}$) for the effects of Barred
Owls on apparent annual survival of adult Northern Spotted Owls on
11 demographic study areas in Washington, Oregon, and California.

Estimates were based on the best $QAIC_c$ model that included the Barred Owl effect.

Study area	$\Delta QAIC_c$	$\hat{\beta}$	$\widehat{SE}$	95% CI Lower	95% CI Upper
Washington					
CLE	3.08	-0.815	1.009	-2.793	1.164
RAI	0.00	-5.330	1.960	-9.190	-1.490
OLY	1.17	-1.216	0.748	-2.682	0.250
Oregon					
COA	9.48	-0.908	0.257	-1.412	-0.405
HJA	2.24	-0.753	0.306	-1.352	-0.153
TYE	9.78	0.062	0.332	-0.588	0.712
KLA	5.21	-0.469	0.655	-1.753	0.815
CAS	4.04	1.657	0.878	-0.062	3.378
California					
NWC	1.98	-1.450	1.079	-3.566	0.666
HUP	1.81	-0.688	1.469	-3.567	2.190
GRD	0.00	-2.234	0.670	-3.547	-0.921
Mean		-1.104	0.514	-2.11	-0.097

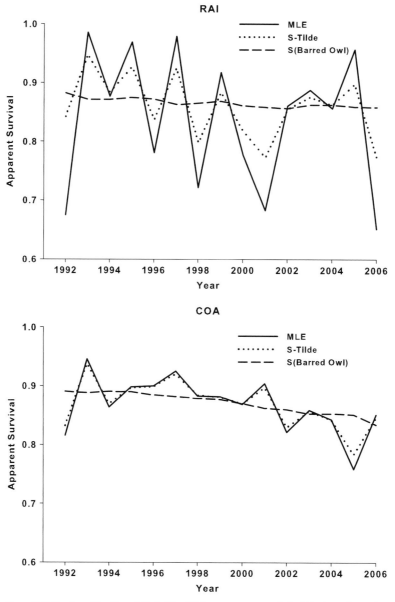

Figure 8. Estimates of the Barred Owl effect (BO) on apparent survival of Northern Spotted Owls. Estimates were generated from the best random effects model [φ(g+t+BO)], plotted with original apparent survival estimates (MLE) and shrinkage estimates (S-tilde) for one study area in Washington (RAI), two study areas in Oregon (CAS, COA), and one study area in California (NWC). Study area effects are represented by g and annual time effects by t.

Owls on survival of Spotted Owls was weak or negligible for CLE, TYE, KLA, and HUP because confidence intervals on regression coefficients overlapped zero (Table 18). With the exception of CLE, the latter areas were all in the southern portion of the range of the Northern Spotted Owl (Fig. 1).

In the meta-analysis of apparent survival, the second best model [φ(g*t) p(g+s+t): RE(g+BO)]

provided strong evidence that the presence of Barred Owls had a negative effect on apparent survival, as the 95% confidence interval on β̂ for the Barred Owl effect did not overlap zero (Table 15; Fig. 8). In addition, the g+BO model ranked higher than the g*BO model, indicating that the BO covariate was important across all study areas in explaining time variation in φ. Thus, there was strong evidence that Barred

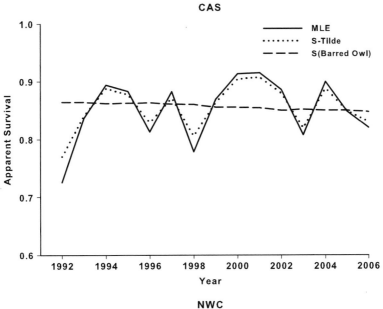

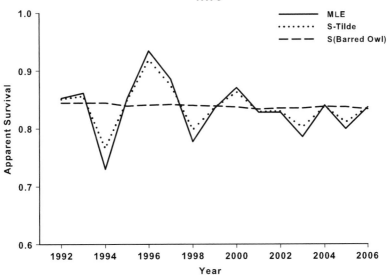

Figure 8. (*continued*)

Owls had a negative effect on apparent survival of Spotted Owls.

Annual Rate of Population Change

Individual Study Areas

We used capture histories of 5,244 banded territorial owls to estimate annual rates of population change (λ) at the 11 study areas. Estimates of goodness-of-fit (χ^2/df) of the capture–recapture data from program RELEASE ranged from 0.69 to 3.02 for individual study areas (Table 11), and the overall estimate of χ^2/df for all of the data combined was 1.61 ($P > 0.10$), indicating good fit of the data to the Cormack–Jolly–Seber model. Estimates of $\hat{c}$ from the median-$\hat{c}$ routine in program MARK ranged from 1.00 to 1.13, indicating little evidence for lack of independence in capture histories (Table 11).

The full sex- and time-specific model φ(s*t) p(s*t) f(s*t) for estimation of λ was not appropriate for most study areas based on model selection with QAIC$_c$. Therefore, we used the time-only model φ(t) p(t) f(t) for estimating λ and temporal process variation for most study areas (Table 19). The only exception was the OLY study area, where there were differences in φ between males and females. Estimates of λ ranged from 0.929 to 0.996 for the 11 study areas and the time span of the estimates ranged from 12 to 16 years (Table 19). There was strong evidence that populations on the CLE, RAI, OLY, COA, HJA, NWC, and GDR study areas declined

during the study, based on 95% confidence intervals for estimates of λ that did not include 1.0 (Table 19, Fig. 9). Estimates of λ for CLE and RAI were especially low, suggesting population declines of 6.3 and 7.1 % per year, respectively (Table 19). Point estimates of λ for the TYE, KLA, CAS, and HUP study areas all indicated declining populations, but had 95% confidence intervals that included 1.0 (Table 19). The weighted mean estimate of λ for all study areas combined was 0.971 (SE = 0.007, 95% CI = 0.960 to 0.983), indicating that the average rate of population decline was 2.9% per year during the study.

TABLE 19

Estimates of λ and temporal process standard deviation ($\hat{\sigma}_{temporal}$) for Northern Spotted Owls on 11 study areas in Washington, Oregon, and California.

Estimates of λ were generated using the best random effects model; estimates of temporal variance are based on random effects models (Means, T, or TT), using time-specific estimates of φ, p, and λ, except where noted.

Study	Years	Model[a]	Derived Å	SE	95% CI Lower	Upper	$\hat{\sigma}_{TEMPORAL}$	95% CI Lower	Upper
Washington									
CLE[b]	1994–2006	[φ(t) p(t) λ(t)]: RE(.)	0.937	0.014	0.910	0.964	0.0000	0.0000	0.0058
RAI	1995–2006	[φ(t) p(t) f(t)]: RE(.)	0.929	0.026	0.877	0.977	0.0048	0.0000	0.0371
OLY	1992–2006	[φ(s*t) p(t) f(t)]: RE(T)	0.957	0.020	0.918	0.997	0.0062	0.0000	0.0332
Oregon									
COA	1994–2006	[φ(t) p(t) f(t)]: RE(T)	0.966	0.011	0.943	0.985	0.0007	0.0000	0.0080
HJA	1992–2006	[φ(t) p(t) f(t)]: RE(TT)	0.977	0.010	0.957	0.996	0.0000	0.0000	0.0042
TYE	1992–2006	[φ(t) p(t) f(t)]: RE(TT)	0.996	0.020	0.957	1.035	0.0012	0.0000	0.0087
KLA	1992–2006	[φ(t) p(t) f(t)]: RE(.)	0.990	0.014	0.962	1.017	0.0019	0.0000	0.0102
CAS	1994–2006	[φ(t) p(t) f(t)]: RE(.)	0.982	0.030	0.923	1.040	0.0105	0.0022	0.0421
California									
NWC	1990–2006	[φ(t) p(t) f(t)]: RE(.)	0.983	0.008	0.968	0.998	0.0000	0.0000	0.0012
HUP	1994–2006	[φ(t) p(t) f(t)]: RE(.)	0.989	0.013	0.963	1.014	0.0000	0.0000	0.0012
GRD	1992–2006	[φ(t) p(t) f(t)]: RE(TT)	0.972	0.012	0.949	0.995	0.0014	0.0000	0.0076
Weighted mean for 8 monitoring areas			0.972	0.006	0.958	0.985			
Weighted mean for 3 non-monitoring areas			0.969	0.016	0.938	1.000			
Weighted mean for all areas			0.971	0.007	0.960	0.983			

[a] Best capture–recapture model structure from analysis of the *a priori* model set. Model notation indicates structure for effects of time (t), linear time trend (T), quadratic time trend (TT), or constant (.), or random effects (RE). For linear and quadratic time trend models, λ was computed using a mean-centered model.

[b] Random effects model using the survival–recruitment parameterization would not run on derived lambdas for CLE. Therefore, we used the survival–lambda *f*-parameterization instead.

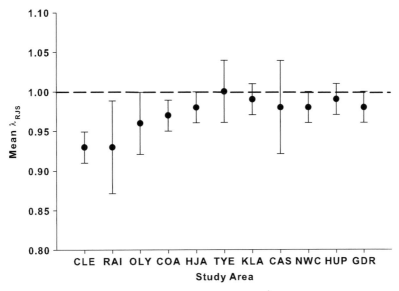

Figure 9. Estimates of mean annual rate of population change ($\hat{\lambda}_{RJS}$), with 95% confidence intervals for Northern Spotted Owls in 11 study areas in Washington, Oregon, and California. Estimates of λ were derived parameters from the recruitment and survival parameterization and the best random effects models based on the best global model [either $f(t)$ $\phi(t)$ $p(t)$ or $f(s*t)$ $\phi(s*t)$ $p(s*t)$], where s and t represent sex and annual time changes, respectively.

Results of the variance components analyses for each study area provided little evidence of temporal process variation in λ for most study areas, relative to the magnitude of sampling variation in estimates (Table 19). Estimates of temporal process variation in λ were highest for the RAI, OLY, CAS, and NWC study areas, but the only study area for which the 95% confidence interval on temporal variation did not include zero was CAS (Table 19).

There was evidence that populations were declining on five of the eight monitoring areas (CLE, OLY, COA, HJA, NWC) based on 95% confidence intervals for λ that did not overlap 1.0. Point estimates of λ for the remainder of the study areas (TYE, KLA, CAS) were less than one, but had confidence intervals that overlapped 1.0, so the evidence for declines on those areas was weak. The weighted mean estimate of λ for the eight monitoring areas was 0.972 (SE = 0.006, 95% CI = 0.958 to 0.985), indicating an estimated decline of 2.8% per year on federal lands within the range of the owl. The weighted mean estimate of λ for the other three study areas (RAI, GDR, HUP) was 0.969 (SE = 0.016,

95% CI = 0.938 to 1.000), indicating an estimated decline of 3.1% per year on those areas.

Estimates of Realized Population Change

Estimates of realized population change indicated that populations in Washington and northern Oregon (OLY, RAI, CLE, COA) declined by 40 to 60% during our study (Fig. 10A, B). There was also evidence that populations on HJA, GDR, and NWC declined during the same period, but the 95% confidence intervals around the estimates of Δ_t on the latter three areas slightly overlapped 1.0 (Fig. 10B, C). Estimates of realized population change for the rest of the study areas (CAS, TYE, KLA, HUP) were all <1.0, but the 95% confidence intervals around the estimates of Δ_t substantially overlapped 1.0. Trends in populations for each of the study areas were variable, and declines, if any, occurred at different times on different areas. For example, the decline on HJA occurred primarily during 1992 to 1993 after a year of high reproductive success in 1992, then the population declined about 10% during

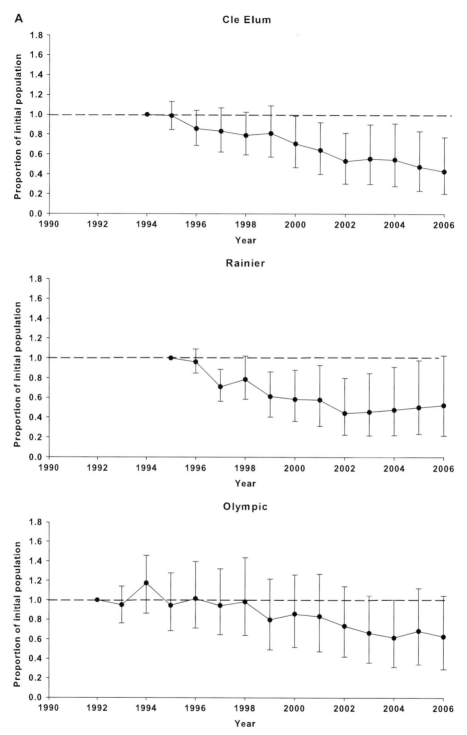

Figure 10. Estimates of realized population change, Δ_t, with 95% confidence intervals for Northern Spotted Owls at three study areas in Washington (A), five study areas in Oregon (B), and three study areas in California (C).

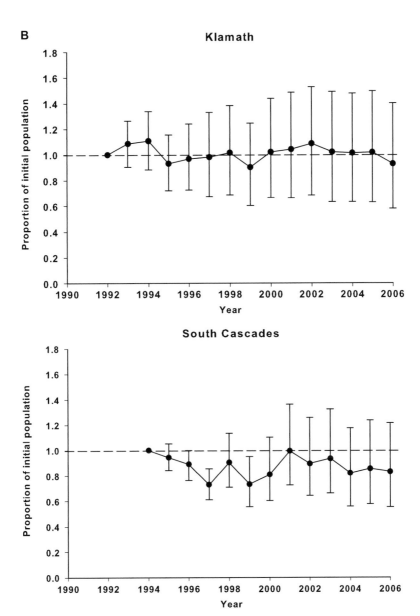

Figure 10. (*continued, for Northern Spotted Owls in Oregon*)

the ensuing decade. In contrast, the decline on COA occurred after 2001 and continued through 2006 (Fig. 10B). Populations in Washington (CLE, RAI, OLY) exhibited a long, gradual decline after the mid-1990s, except that the population on RAI actually increased slightly after 2002 (Fig. 10A). Consequently, there was no evidence for synchrony in timing of population declines among the 11 study areas.

Meta-analysis of Annual Rate of Population Change

Estimates of goodness-of-fit from program RELEASE for individual study areas (Table 11) indicated good fit of the data to the Cormack–Jolly–Seber model for all study areas. In addition, the mean estimate of median-$\hat{c}$ from program MARK was 1.06 with a range of 1.0 to 1.17,

B *continued*

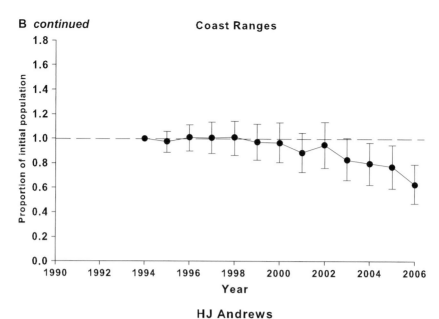

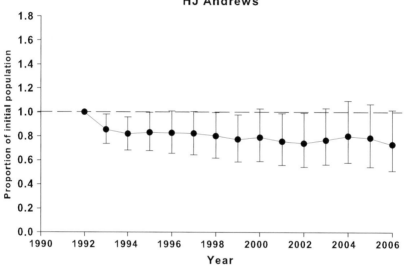

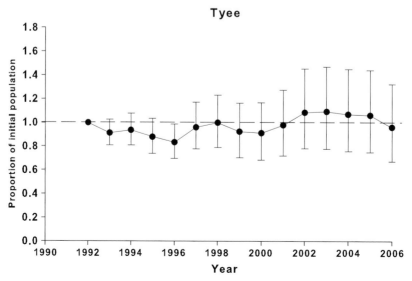

Figure 10. (*continued, for Northern Spotted Owls in Oregon*)

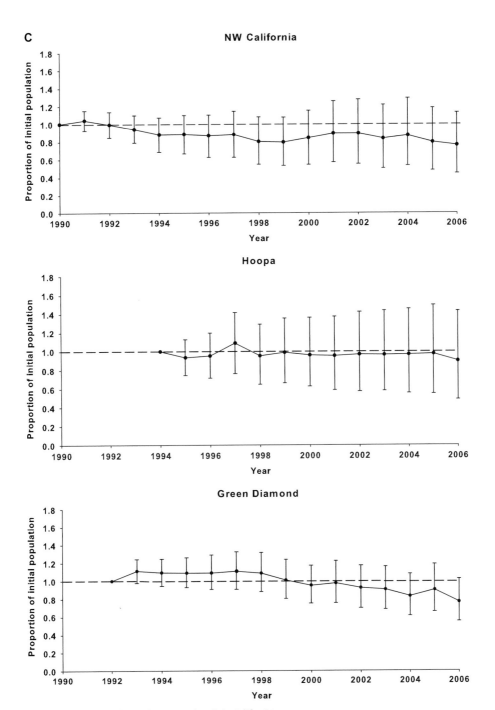

Figure 10. (*continued, for Northern Spotted Owls in California*)

indicating little evidence for overdispersion (i.e., lack of independence) in the capture–recapture data. As a result, we did not use $\hat{c}$ to adjust model selection to QAIC$_c$ or inflate variance estimates of parameters.

The best *a priori* model in the meta-analysis of λ was RE (random effects) model φ(ECO) *f*(ECO), which indicated evidence of an effect of ecoregion on φ and *f* (Table 20). Two competing random effects models had ΔAIC$_c$ values

TABLE 20

Model selection results from meta-analysis of λ for Northern Spotted Owls in Washington, Oregon, and California.

Model[a]	K	Deviance	AIC$_c$	ΔAIC$_c$	w_i
[φ(g*t) p(g*t) f(g*t)]: RE φ(ECO + BO) f(ECO)*	500.85	17,924.51	60,812.29		
[φ(g*t) p(g*t) f(g*t)): RE φ(ECO + BO) f(ECO + BO)*	501.01	17,924.65	60,812.76		
[φ(g*t) p(g*t) f(g*t)]: RE φ(ECO) f(ECO)	501.44	17,924.22	60,813.25	0.00	0.302
[φ(g*t) p(g*t) f(g*t)]: RE φ(ECO) f(ECO*BO)*	501.89	17,923.45	60,813.43		
[φ(g*t) p(g*t) f(g*t)]: RE φ(ECO + BO) f(ECO)*	501.53	17,924.33	60,813.54		
[φ(g*t) p(g*t) f(g*t)]: RE φ(g + BO) f(BO)	502.32	17,922.77	60,813.64	0.39	0.248
[φ(g*t) p(g*t) f(g*t)]: RE φ(ECO) f(ECO + BO)*	501.60	17,924.37	60,813.73		
[φ(g*t) p(g*t) f(g*t)]: RE φ(ECO) f(OWN + ECO)	501.94	17,924.41	60,814.49	1.24	0.162
[φ(g*t) p(g*t) f(g*t)]: RE φ(ECO*BO) f(ECO*BO)*	502.36	17,923.74	60,814.69		
[φ(g*t) p(g*t) f(g*t)]: RE φ(g + BO) f(g + BO)	502.63	17,925.46	60,816.98	3.73	0.047
[φ(g*t) p(g*t) f(g*t)]: RE φ(g + BO) f(g*BO)	503.37	17,924.01	60,817.08	3.83	0.044
[φ(g*t) p(g*t) f(g*t)]: RE φ(g) f(g)	503.35	17,925.06	60,818.09	4.84	0.027
[φ(g*t) p(g*t) f(g*t)]: RE φ(g) f(g + TT)	503.76	17,924.24	60,818.14	4.89	0.026
[φ(g*t) p(g*t) f(g*t)]: RE φ(g + PDSI) f(g + ENP + ENT)	503.73	17,924.59	60,818.43	5.18	0.023
[φ(g*t) p(g*t) f(g*t)]: RE φ(g) f(g + T)	503.62	17,924.93	60,818.54	5.29	0.021
[φ(g*t) p(g*t) f(g*t)]: RE φ(g + PDSI) f(g + LNP)	503.79	17,924.85	60,818.82	5.56	0.019
[φ(g*t) p(g*t) f(g*t)]: RE φ(g + PDSI) f(g + PDSI)	503.78	17,924.91	60,818.85	5.59	0.018
[φ(g*t) p(g*t) f(g*t)]: RE φ(g + PDSI) f(g + SO + PDO)	503.83	17,924.89	60,818.94	5.69	0.018
[φ(g*t) p(g*t) f(g*t)]: RE φ(g) f(g*T)	505.03	17,922.98	60,819.55	6.30	0.013
[φ(g*t) p(g*t) f(g*t)]: RE φ(g*T) f(g)	504.13	17,924.99	60,819.66	6.41	0.012
[φ(g*t) p(g*t) f(g + t)]	395.00	18,154.00	60,820.54	7.29	0.008
[φ(g*t) p(g*t) f(g*t)]: RE φ(g + PDSI) f(g*LNP)	505.93	17,923.27	60,821.73	8.48	0.004
[φ(g*t) p(g*t) f(g*t)]: RE φ(g + PDSI) f(g*PDSI)	505.89	17,923.37	60,821.76	8.51	0.004
[φ(g*t) p(g*t) f(g*t)]: RE φ(g) f(g*TT)	508.04	17,919.98	60,822.88	9.63	0.002
[φ(g*t) p(g*t) f(g*t)]: RE φ(g + PDSI) f(g*ENP + g*ENT)	508.44	17,921.51	60,825.24	11.99	0.001
[φ(g*t) p(g*t) f(g*t)]: RE φ(g + PDSI) f(g*SOI + g*PDO)	508.52	17,922.20	60,826.11	12.86	0.000
[φ(g*t) p(g*t) f(g*t)]: RE φ(g*HAB2^k) f(g + HAB2 + HAB3)	518.79	17,914.06	60,839.59	26.33	0.000
[φ(g*t) p(g*t) f(g*t)]: RE φ(g*HAB2) f(g*HAB3)	520.17	17,912.94	60,841.36	28.11	0.000
[φ(g*t) p(g*t) f(g*t)]: RE φ(g)	524.84	17,904.03	60,842.29	29.04	0.000
[φ(g*t) p(g*t) f(g*t)]: RE φ(g*HAB2) f(g*HAB2 + g*HAB3)	521.38	17,911.71	60,842.68	29.43	0.000
[φ(g*t) p(g*t) f(g*t)]: RE φ(g*TT)	527.03	17,903.49	60,846.36	33.11	0.000
[φ(g*t) p(g*t) f(g*t)]: RE φ(ECO)	527.08	17,904.21	60,847.17	33.92	0.000
[φ(g*t) p(g*t) f(g*t)]: RE φ(g + BO)	527.35	17,904.03	60,847.56	34.31	0.000
[φ(g*t) p(g*t) f(g*t)]: RE φ(g*HAB2) f(g + HAB2)	527.19	17,907.03	60,850.23	36.98	0.000
[φ(g*t) p(g*t) f(g*t)]: RE φ(g + PDSI)	528.95	17,904.03	60,850.95	37.70	0.000

TABLE 20 (continued)

TABLE 20 (CONTINUED)

Model[a]	K	Deviance	AIC_c	ΔAIC_c	w_i
[φ(g*t) p(g*t) f(g*t)]: RE φ(BO)	529.32	17,904.28	60,851.96	38.71	0.000
[φ(g*t) p(g*t) f(g*t)]: RE φ(OWN + ECO)	529.40	17,904.12	60,851.97	38.72	0.000
(φ(g*t) p(g*t) f(g*t)): RE φ(LAT)	529.38	17,904.29	60,852.10	38.85	0.000
[φ(g*t) p(g*t) f(g*t)]: RE φ(g + T)	529.60	17,904.03	60,852.30	39.04	0.000
[φ(g*t) p(g*t) f(g*t)]: RE φ(OWN)	529.62	17,904.24	60,852.56	39.31	0.000
[φ(g*t) p(g*t) f(g*t)]: RE φ(g*PDSI)	530.40	17,904.10	60,854.05	40.80	0.000
[φ(g*t) p(g*t) f(g*t)]: RE φ(g + SOI + PDO)	529.80	17,905.65	60,854.35	41.09	0.000
[φ(g*t) p(g*t) f(g*t)]: RE φ(g*T)	530.78	17,903.78	60,854.54	41.28	0.000
[φ(g*t) p(g*t) f(g*t)]: RE φ(g*BO)	530.80	17,903.91	60,854.72	41.46	0.000
[φ(g*t) p(g*t) f(g*t)]: RE φ(g + ENP + ENT)	530.11	17,905.61	60,854.95	41.70	0.000
[φ(g*t) p(g*t) f(g*t)]: RE φ(g*SOI + g*PDO)	531.57	17,903.55	60,855.99	42.73	0.000
[φ(g*t) p(g*t) f(g*t)]: RE φ(g*HAB2)	531.50	17,904.29	60,856.57	43.32	0.000
[φ(g*t) p(g*t) f(g*t)]: RE φ(g*ENP + g*ENT)	531.84	17,905.15	60,858.14	44.89	0.000
[φ(g*t) p(g*t) f(g*t)]:RE φ(g + HAB2)	534.12	17,902.83	60,860.63	47.38	0.000
[φ(g*t) p(g*t) f(g*t)]: RE φ(g + TT)	529.39	17,912.96	60,860.79	47.54	0.000
φ(g*t) p(g*t) f(g*t)	542.00	17,922.47	60,896.89	83.64	0.000

NOTE: Model form was the survival and recruitment parameterization. Notation for random effects (RE) models includes the general model on which the random effects model is based (g = study area, t = time varying). Models ending with asterisks were developed *a posteriori* after seeing the results of the original modeling. Inferences were based on the models in the original *a priori* model set.

[a] Model notation indicates structure for study area (g), time (t), linear time trend (T), quadratic time trend (TT), ecoregion (ECO), land ownership (OWN), proportion of territories with Barred Owl detections (BO), early nesting season precipitation (ENP), early nesting season temperature (ENT), late nesting season precipitation (LNP), late nesting season temperature (LNT), Palmer Drought Severity Index (PDSI), percent cover of suitable owl habitat within a 2.4 km radius of owl activity centers (HAB2), percent cover of suitable owl habitat within 23 km of owl activity centers, minus the area within 2.4 km of owl activity centers (HAB3), latitude (LAT), Southern Oscillation Index (SOI), and Pacific Decadal Oscillation (PDO).

<2.0, one of which indicated evidence of a Barred Owl effect on φ and *f* [φ(g+BO) *f*(BO)], and one [φ(ECO) *f*(ECO+OWN)] that indicated differences in recruitment among different land ownership categories (Table 20). The 95% confidence interval for the effects of ownership on *f* in the latter model included zero, indicating little evidence of an effect of ownership on recruitment (Table 21). Therefore, model selection results for the top two models [φ(ECO) *f*(ECO) and φ(g+BO) *f*(BO)] indicated the most support for models that included Barred Owls (BO) and ecoregions (ECO). Estimates of apparent survival from the best *a priori* model were highest for the Oregon Coast Douglas-fir ecoregion and lowest for the Washington Mixed-conifer ecoregion (Fig. 11). Recruitment was highest in the Oregon/California Mixed-conifer ecoregion

($\hat{f}$ = 0.145, SE = 0.020), but similar among the other ecoregions (Fig. 11). The low estimates of λ for the Washington Douglas-fir and Washington Mixed-conifer ecoregions were a result of both low apparent survival and low recruitment. In contrast, the Oregon/California Mixed-conifer region had the highest estimate of λ, which was a result of high recruitment and intermediate survival rates. Values of φ, *f*, and λ were intermediate for the other ecoregions.

Slope coefficients for the Barred Owl effect in the random effects (RE) model φ(g+BO) p(g*t) *f*(BO) were negatively associated with apparent survival and recruitment, although the 95% confidence interval for the effect of Barred Owls on recruitment included zero (Table 21). There was some evidence for differences in apparent survival among different land ownership categories

TABLE 21

Coefficient estimates (β̂) for the best models that included effects of Barred Owls, land ownership, climate, habitat, or latitude in the meta-analysis of λ for 11 study areas in Washington, Oregon, and California.

	Survival				Recruitment			
			95% CI				95% CI	
Covariate[a]	β̂	$\widehat{SE}$	Lower	Upper	β̂	$\widehat{SE}$	Lower	Upper
BO	-0.116	0.043	-0.200	-0.032	-0.023	0.037	-0.096	0.050
Ownership								
Federal (intercept)	0.869	0.020	0.829	0.908	0.098	0.020	0.058	0.137
Non-federal	0.023	0.022	-0.020	0.067	-0.027	0.023	-0.073	0.019
Mixed	0.002	0.013	-0.023	0.027	-0.002	0.013	-0.028	0.024
Climate								
ENP	0.007	0.007	-0.006	0.021	0.012	0.007	-0.002	0.026
ENT	0.000	0.000	0.000	0.000	0.000	0.000	-0.001	0.000
LNP	na				0.000	0.001	-0.002	0.002
PDSI	0.002	0.002	-0.002	0.006	-0.001	0.002	-0.006	0.004
SOI	0.007	0.008	-0.009	0.023	-0.010	0.009	-0.027	0.007
PDO	0.017	0.008	0.000	0.033	-0.001	0.009	-0.018	0.017
Habitat								
HAB2					0.559	0.285	0.001	1.117
HAB3					-0.688	0.303	-1.282	-0.093
HAB2-CAS	0.602	1.291	-1.928	3.131				
HAB2-HJA	6.851	4.117	-1.218	14.921				
HAB2-KLA	-0.477	1.060	-2.554	1.600				
HAB2-OLY	-3.749	16.270	-35.638	28.141				
HAB2-RAI	-0.470	0.342	-1.141	0.202				
HAB2-CLE	1.143	1.004	-0.824	3.111				
HAB2-COA	1.155	0.922	-0.651	2.962				
HAB2-TYE	0.763	0.671	-0.554	2.079				
LAT	-0.002	0.002	-0.007	0.002				

[a] Covariates included proportion of territories with Barred Owl detections (BO), early nesting season precipitation (ENP), early nesting season temperature (ENT), late nesting season precipitation (LNP), Palmer Drought Severity Index (PDSI), Southern Oscillation Index(SOI), Pacific Decadal Oscillation (PDO), percent cover of suitable owl habitat within a 2.4-km radius of owl activity centers (HAB2), forest habitat in the ring between HAB2 and a circle defined by the median natal dispersal distance (23 km) (HAB3), and latitude (LAT).

but the differences were minor, and the best model that included the ownership covariate ranked far below the top model ($\Delta AIC_c = 38.72$; Table 20). There was no evidence that latitude or habitat within the study area (HAB2) had an effect on apparent survival, but there was evidence that apparent survival was positively related to the Pacific Decadal Oscillation (β̂ = 0.017, 95% CI = 0.0002 to 0.033; Table 21), which was consistent with our prediction. Other

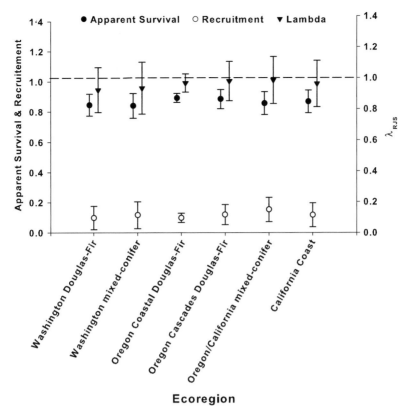

Figure 11. Point estimates and 95% confidence limits of apparent survival, recruitment, and λ of Northern Spotted Owls in different ecoregions based on the best *a priori* model from the meta-analysis of 11 study areas [RE φ(ECO) *f*(ECO)].

climate covariates explained little of the variation in apparent survival rates (Table 21). Lack of evidence of an effect of habitat and weather on apparent survival may represent a true absence of an effect, but we cannot rule out the possibility that the lack of an effect resulted from the covariates being computed at too coarse a scale, or because the definitions we used to map habitat did not accurately reflect suitable habitat.

Examination of the relationship between recruitment and ownership indicated a weak effect, with slightly higher recruitment on federal lands ($\hat{\beta}$ = 0.098, 95% CI = 0.058 to 0.137) than on mixed federal–private and private lands (Table 21). Although habitat covariates did not appear in any of the top models in the meta-analysis of λ, examination of the best models that included habitat covariates provided evidence that the percent of the study area covered by suitable owl habitat had a positive effect on

recruitment (covariate HAB2 in Table 21). In contrast, recruitment was negatively related to the percent of the area surrounding the study area that was covered by suitable owl habitat (covariate HAB3 in Table 21). Our results may reflect an interaction or synergistic relationship between recruitment and the percent cover of suitable owl habitat within versus surrounding the study areas on federal lands compared to other land ownerships. We did not include such models in our *a priori* model set, so these relationships should be investigated in more detail in future analyses. There was no evidence that recruitment was influenced by any of our weather or climate covariates as all 95% confidence intervals for these covariates included zero (Table 21).

Plots of year-specific estimates of φ_t and f_t indicated considerable temporal and spatial variation, which produced high temporal and spatial variation in λ (Fig. 12). For example, all

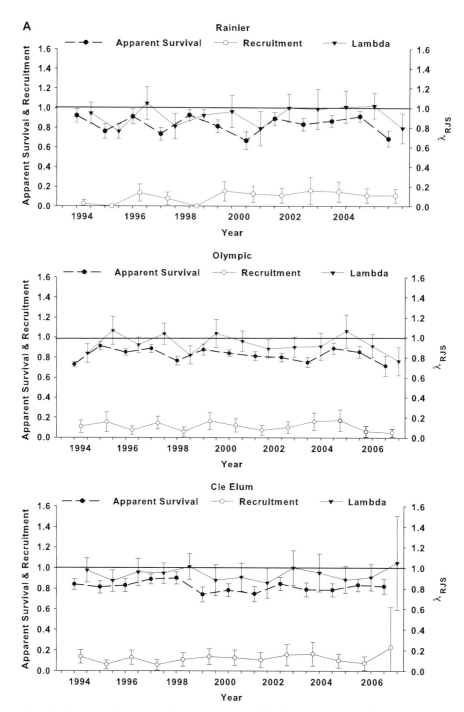

Figure 12. Estimates of apparent survival, recruitment, and λ of Northern Spotted Owls based on the most general model [(g*t) f(g*t)] from the meta-analysis of three study areas in Washington (A), five study areas in Oregon (B), and three study areas in California (C). Vertical bars indicate 95% confidence limits, and g and t represent study area and annual time effects, respectively.

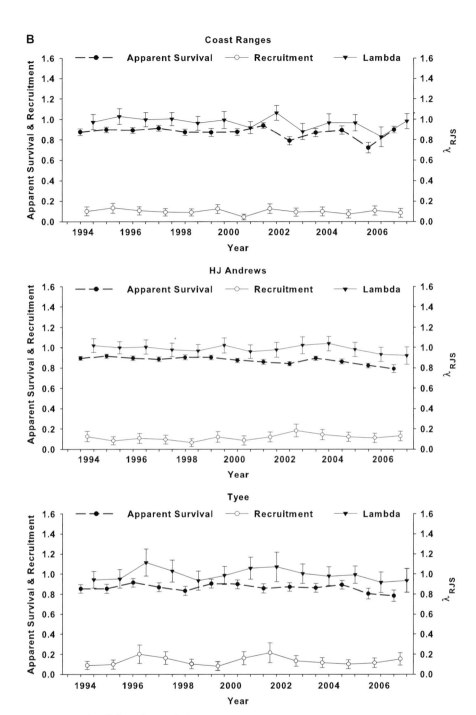

Figure 12. (*continued, for study areas in Oregon*)

B *continued*

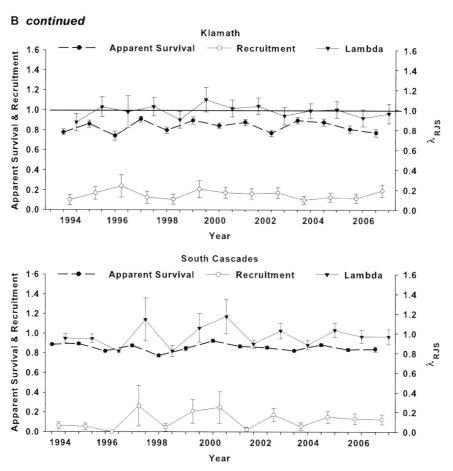

Figure 12. (*continued, for study areas in Oregon*)

three parameters (φ_t, f_t, λ) exhibited considerable variation in Washington where owl populations were declining the most (Fig. 12A), but less variation in most of the other study areas. Temporal variation in φ_t was paralleled by temporal variation in λ_t for most study areas (OLY, CLE, COA, HJA, TYE, KLA, NWC, HUP, GDR), suggesting that changes in λ_t were influenced primarily by changes in survival. However, this pattern was not as evident for RAI and CAS during all years, and there was evidence that recruitment had a substantial influence on λ_t in those two areas, particularly during years when λ_t increased noticeably. In addition, estimated recruitment was essentially zero in some years on the RAI, OLY, and CAS study areas, which resulted in noticeable declines in λ_t, since φ was always <1.0. Overall, the high temporal variation in the annual rate of population change of

Spotted Owls was closely associated with apparent survival rates in most cases and with recruitment in a few cases.

DISCUSSION

The Northern Spotted Owl has been the "poster child" for conservation of old-growth and mature forests in the Pacific Northwest and has served as an "umbrella species" (Roberge and Angelstam 2004) for conservation of other species associated with old forests (USDA Forest Service and USDI Bureau of Land Management 1994). As a result, numerous conservation plans have addressed the habitat needs of Spotted Owls on federal lands. In conjunction with the listing of the subspecies as threatened in 1990, the Interagency Scientific Committee (ISC) developed and published the first comprehensive conservation

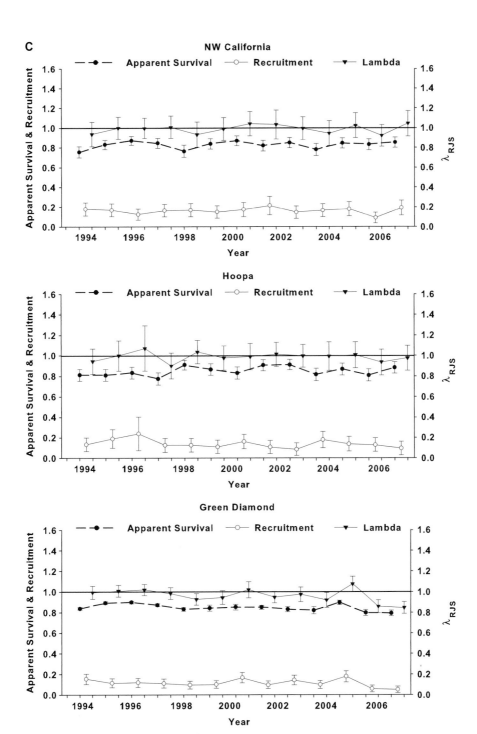

Figure 12. (*continued, for study areas in California*)

plan for the Northern Spotted Owl (Thomas et al. 1990). The ISC plan called for the conservation of an unprecedented amount of old forest in large reserves that were spaced within 19.2 km of each other and large enough to support 20 to 25 pairs of territorial owls. The ISC conservation strategy was the framework, with minor modifications, for the first draft final recovery plan for the Northern Spotted Owl (USDI Fish and Wildlife Service 1992), and also served as a model for the network of old forest reserves that eventually became the Northwest Forest Plan for management of all federal lands within the geographic range of the subspecies (USDA Forest Service and USDI Bureau of Land Management 1994).

The Northwest Forest Plan served as the *de facto* recovery plan for the Northern Spotted Owl for approximately 14 years during which time there was no approved recovery plan for the owl. The situation changed in 2008, when the U.S. Fish and Wildlife Service published a final recovery plan for the Northern Spotted Owl (USDI Fish and Wildlife Service 2008). The 2008 recovery plan included a much-reduced network of old forest reserves compared to the Northwest Forest Plan, and the approach laid out in the recovery plan was criticized by three professional societies concerned about the recovery of the owl (e.g., Wildlife Society 2008). The U.S. Department of Justice subsequently declined to defend the 2008 recovery plan, and it was remanded to the Fish and Wildlife Service with instructions that they address the deficiencies noted by their critics. At this writing, the Fish and Wildlife Service is working on a revision of the 2008 plan, but the situation is still unresolved.

Because the Northern Spotted Owl is federally listed as "Threatened" under the Endangered Species Act (USDI Fish and Wildlife Service 1990), and is the focus of many forest management practices that have been implemented in recent years in the Pacific Northwest, results of our study will be of interest to a number of stakeholders, including state and federal government agencies, conservation groups, private industry, and the public. Consequently, it is important to ask: What is our frame of reference and what

kind of inferences can we make from the results of our study? From a statistical standpoint, a formal inference can be made from the sample of marked and recaptured owls to the population of owls in the study areas in which the marked owls were located. Our 11 study areas covered a large portion of the subspecies' geographic range and included substantial variation in latitude, elevation, and land ownership (Appendix A), but they were not selected randomly. Consequently, the results of our analyses cannot be considered representative of demographic trends of Northern Spotted Owls throughout their entire range. For example, there were no study areas in the extensive areas of state and private lands in northwestern Oregon and southwestern Washington or in the California Cascades. However, we believe that our results are representative of most populations of Northern Spotted Owls in the Pacific Northwest that are on federal lands or in areas of mixed federal and private ownership. We do not think that our results can be used to assess demographic trends of Spotted Owls on non-federal lands because the two study areas in our sample that were entirely on non-federal lands (GDR, HUP) were atypical. Both the Green Diamond Resource Company and the Hoopa Tribe managed their lands to protect known Spotted Owl nest areas and to maintain at least part of their lands in suitable foraging habitat for Spotted Owls. Such practices are not universal on private and state lands. If anything, our results probably depict an optimistic view of the overall population status of the Northern Spotted Owl.

This study is the fifth meta-analysis of demographic data from Northern Spotted Owls (Anderson and Burnham 1992, Burnham et al. 1996, Franklin et al. 1999, Anthony et al. 2006); however, only two of these efforts were published as refereed journal articles (Burnham et al. 1996, Anthony et al. 2006). The other articles are not readily available, so we will concentrate our discussion on the two published articles. The second meta-analysis of demographic rates of Northern Spotted Owls was conducted in 1993 and included 11 study areas (Burnham et al. 1996, Forsman et al. 1996a). The three major findings of the second analysis were: (1) Fecundity rates varied among years and ages of owls, with no increasing or decreasing trend over time; (2) survival rates were dependent on age and there was a decreasing trend in adult female survival; (3) the annual rate of population change (λ_{PM}) was <1.0 for 10 of 11 areas examined, and the estimated average rate of population decline was 4.5% per year (Burnham et al. 1996). Results of the first three meta-analyses of demography of Northern Spotted Owls were critiqued by Raphael et al. (1996) and Boyce et al. (2005), who questioned the estimates of annual rate of population change from Leslie matrix models (λ_{PM}), primarily because estimates of juvenile survival from capture–recapture methods were biased by permanent emigration during natal dispersal. Anthony et al. (2006) avoided this problem by using the Pradel (1996) model, which estimates the annual finite rate of population change (λ_{RJS}) of territorial owls without inclusion of juvenile survival rates. In addition, the Pradel (1996) model treats losses due to emigration and mortality and gains due to recruitment and survival in a symmetric way, so it is less subject to biases in the estimate of λ. For more information on this topic, see Anthony et al. (2006), and for a review of the differences between λ_{PM} and λ_{RJS}, see Sandercock and Beissinger (2002).

The most important findings in the Anthony et al. (2006) report were: (1) Fecundity was relatively stable among the 14 study areas examined, (2) survival rates were declining on 5 of the 14 areas, and (3) populations were declining on 9 of 13 study areas for which there was adequate data to estimate λ. The mean λ for the 13 areas was 0.963, which indicated that populations were declining 3.7% annually during the study (Anthony et al. 2006:34). The reasons for declines in Spotted Owl populations in their study were not readily apparent. Therefore, Anthony et al. (2006) recommended the use of additional covariates in future analyses to evaluate the possible influence of Barred Owls, weather, habitat, and reproduction on vital rates and population trends of Spotted Owls.

Fecundity

The results from our analysis of fecundity were consistent with previous analyses in that we found substantial annual variation in fecundity on individual study areas and a biennial cycle of high fecundity in even-numbered years and low fecundity in odd-numbered years (Burnham et al. 1996, Anthony et al. 2006). The cause of this synchronization remains unknown. One hypothesis for alternate year breeding in long-lived species that require many months to produce a single brood is that reproduction every year is physically impossible because of the large investment of time and energy required to produce a single brood. A hypothesis of intermittent breeding makes sense for some long-lived alternate year breeders such as Albatross (*Diomedea exulans*, *Phoebetria fusca*, *P. palpebrata*), which have to travel huge distances for many months in order to provision a single young (Tickell 1968, Weimerskirch et al. 1987). Although Spotted Owls also invest many months to produce a single brood (Mar–Aug), there is considerable variation among individuals regarding the alternate year pattern of breeding. In some of our study areas, the majority of owls nested every other year, but there were a few pairs that nested in nearly all years, and there were many that did not follow a predictable pattern. We conclude that breeding in the Spotted Owl is a complex interaction between age, prey abundance, weather, individual variation, and territory quality. However, none of these factors are known to fluctuate on a two-year cycle on our study areas, and prey cycles observed in other studies generally suggest cycles of three years or longer (Korpimaki 1992). Another hypothesis is that the likelihood of breeding is somehow influenced by the molt, which in Spotted Owls is characterized by an alternate year molt of the remiges and rectrices (Forsman 1981). The molt hypothesis seems unlikely, however, as no evidence indicates that the molt was synchronized within the owl populations. The molt hypothesis also does not explain the fact that the even–odd year effect became less evident in the last five years of our study.

Another consistent effect across study areas was variation in fecundity by age class. Fecundity was higher for adults than for 1-yr-olds, and 2-yr-olds were intermediate. A pattern of increasing fecundity with age is typical in birds (Clutton-Brock 1988, Saether 1990), and, in the case of territorial predators like Spotted Owls, probably reflects increased experience and familiarity with a territory and a long-term mate. Spotted Owls in the 1- and 2-yr-old age classes typically comprised <10% of the territorial population, so they contributed little to annual reproduction compared to adults. Age effects were not unexpected and have been well documented in previous studies of Northern Spotted Owls (Burnham et al. 1996, Anthony et al. 2006), California Spotted Owls (*S. o. occidentalis*; Blakesley et al. 2001), and Mexican Spotted Owls (*S. o. lucida*; Seamans et al. 1999, 2001), and are typical of long-lived birds in general (Newton 1989). Compared to the previous meta-analysis of Northern Spotted Owls (Anthony et al. 2006), the addition of five years of data resulted in slightly lower mean fecundity across study areas for adults ($\bar{x} = 0.340$ vs. 0.372) and 2-yr-olds ($\bar{x} = 0.195$ vs. 0.208), but slightly higher fecundity for 1-yr-olds ($\bar{x} = 0.103$ vs. 0.074). However, our fecundity estimates were still well within the range of values reported on the same study areas during 1985 to 1994 (Burnham et al. 1996). Our results suggested that fecundity was declining in five areas (CLE, KLA, CAS, NWC, GDR), stable in three areas (OLY, TYE, HUP), and increasing in three areas (RAI, COA, HJA). Given the variation in trends among study areas, it was not surprising that the best or competitive models in the meta-analyses of fecundity did not include time trends in fecundity. Our results also were in contrast to a previous analysis in which fecundity appeared to be declining in only two study areas in Washington (Anthony et al. 2006).

In our analysis of individual study areas, there was evidence that the proportion of Spotted Owl territories with detections of Barred Owls had a negative effect on fecundity in four study areas (COA, KLA, CAS, GDR) and an unexpected positive effect on fecundity in one area (HJA). The high frequency of study areas with little evidence of effects of Barred Owls on fecundity did not support our hypothesis of competitive interactions, but findings of negative effects of Barred Owls on some study areas were in contrast to Anthony et al. (2006), who found little evidence of a Barred Owl effect on fecundity. In addition, there was weak evidence for a negative effect of Barred Owls on fecundity in both of our meta-analyses of fecundity. One explanation for the relatively weak effect of Barred Owls on fecundity in studies such as ours is that Barred Owls may simply displace Spotted Owls from their territories. When this happens, Spotted Owls enter the non-territorial population, where they are non-breeders and less detectable using the calling surveys used to sample territorial owls (Kelly 2001). Under this scenario, Spotted Owls that are not displaced may continue to breed at levels similar to historic levels, but the net effect of Barred Owls on fecundity is to reduce the total number of young Spotted Owls produced. Displacement of territorial Spotted Owls by Barred Owls may explain seemingly counterintuitive results such as the positive beta associated with the BO covariate in the analysis of fecundity on the HJA study area. In this situation, the Spotted Owls that are monitored are mostly the ones not displaced by Barred Owls, and are likely to be the oldest and most experienced owls. In addition, detections of Barred Owls were more frequent in our study areas in Washington and Oregon, so we did not expect the effects of Barred Owls to be as strong in California.

While climate and weather covariates explained little of the variation in fecundity in the meta-analysis, there was some support for climate or weather effects in the analyses of individual study areas. For example, there was evidence that low temperatures during the early nesting season had negative effects on fecundity in three study areas (RAI, COA, CAS) and had a positive effect on fecundity in one area (HUP). There was also evidence that high precipitation during the early nesting season had negative effects on fecundity in three study areas (CLE, KLA, NWC). Based on a territory-specific study of Spotted Owls on the TYE study area, Olson et al. (2004) also found evidence for a negative effect of precipitation during the early nesting season on fecundity in 1988 to 1999. Cold, wet weather during the incubation, brooding, and early fledgling stages has been reported to be a direct cause of egg and chick mortality through chilling and exposure in Peregrine Falcons (*Falco peregrinus*; Olsen and Olsen 1989, Bradley et al. 1997) and Australian Brown Falcons (*F. berigora*; McDonald et al. 2004). We also observed mortality in cases where recently fledged owlets died from exposure during unseasonal periods of cold, snowy weather in late May or early June. However, it is unclear if the effect of precipitation on fecundity is due primarily to direct loss of eggs or juveniles from exposure, effects on prey abundance or availability, or reduced foraging efficiency of adults (Franklin et al. 2000). Most likely, the effect is due to a combination of all of these factors. Studies of corticosterone levels show that inclement weather can lead to increased stress among adult birds in Dark-eyed Juncos (*Junco hyemalis*; Rogers et al. 1983), Storm Petrels (*Pelecanoides urinatrix*; Smith et al. 1994), Lapland Longspurs (*Calcarius lapponicus*; Astheimer et al. 1995), White-crowned Sparrows (*Zonotrichia leucophrys*; Wingfield et al. 1983), and male Song Sparrows (*Melospiza melodia*; Wingfield 1985). However, some studies also suggest that only unusually severe weather actually results in stress levels high enough to cause birds to forego nesting or to fail after starting to nest (Romero et al. 2000).

Dugger et al. (2005) suggested that a negative relationship between fecundity of Spotted Owls and mean precipitation in the previous winter could reflect climate effects on prey abundance and/or availability. Few studies have linked abundance or availability of Spotted Owl prey to weather conditions, but Lehmkuhl et al. (2006b) reported that annual survival of northern flying squirrels (*Glaucomys sabrinus*) was negatively associated with snow depth. Fecundity of Spotted Owls could also be influenced by prey abundance. Rosenberg et al. (2003) reported a positive correlation between fecundity of Northern Spotted Owls and abundance of deer mice (*Peromyscus maniculatus*) during the nesting season over an eight-year period on the HJA study area. However, deer mice were not the most important prey in the diet on the HJA study area (<10% of prey numbers), so it was unclear if the correlation between owl fecundity and deer mouse numbers was a causal relationship. Similarly, Ward and Block (1995) documented a year of high reproduction by Mexican Spotted Owls (*S. o. lucida*) that occurred in conjunction with an eruption of white-footed mice (*P. leucopus*) in southern New Mexico. Although the data are limited for Spotted Owls, annual variation in prey abundance has strong effects on fecundity of most raptors in northern latitudes, including such diverse species as Tengmalm's Owl (*Aegolius funereus*; Korpimäki 1992, Hakkarainen et al. 1997), Golden Eagle (*Aquila chrysaetos*; Steenhof et al. 1997), Great-horned Owl (*Bubo virginianus*; Rohner 1996), and Northern Goshawk (*Accipiter gentilis*; Salafsky et al. 2005). We suspect, therefore, that we will continue to have difficulty modeling annual variation in fecundity of Northern Spotted Owls without long-term information on the abundance of prey that make up the majority of their diet, especially flying squirrels, woodrats (*Neotoma* spp.), red-backed voles (*Myodes* spp.), deer mice, tree voles (*Arborimus* spp.), and lagomorphs (*Lepus americanus*, *Sylvilagus* spp.).

In Washington and Oregon, the habitat covariate was included in either a top fecundity model or a competitive model in seven of the eight study areas. There was strong evidence for a positive effect of the amount of habitat on fecundity in four study areas (COA, HJA, TYE, CAS), and a negative effect of habitat on fecundity in one area (KLA). We cannot discount the possibility that the absence of a strong effect of habitat on fecundity in all study areas was because our habitat covariate was too simplistic. Other habitat features such as the amount of edge, mean patch size, or amount of interior forest habitat may be important to Spotted Owls (Franklin et al. 2000, Olson et al. 2004, Dugger et al. 2005), and these variables were not readily available for all of our study areas. Also, in a previous territory-specific study on the NWC study area, Franklin et al. (2000) found that fecundity of Spotted Owls was negatively associated with the amount of interior forest and positively associated with the amount of edge, whereas adult survival was positively associated with the amount of interior old-growth forest and with the amount of edge. Based on these findings, Franklin et al. (2000) postulated that "habitat fitness" for Spotted Owls was greatest in areas that included large amounts of interior mature and old-growth forest, but with considerable amounts of edge as well. However, evidence for a positive effect of edge on fecundity of Spotted Owls is not consistent across the range of the subspecies. For example, Dugger et al. (2005) found a positive relationship between fecundity and the percent cover of old forest within a 730-m-radius circle of Spotted Owl activity centers in southern Oregon but found no evidence that fecundity was positively associated with the amount of edge. Whether spatially explicit covariates such as the amount of edge or amount of interior old forest could be useful or meaningful in a study-area–specific analysis or in a meta-analysis of multiple study areas is questionable but should be explored.

The meta-analysis of adult fecundity also indicated differences among ecoregions and substantial annual variability with no apparent time trend. Our results were virtually identical to those reported by Anthony et al. (2006), including the high fecundity of Spotted Owls in the Washington Mixed-conifer ecoregion compared to all other regions. There was also some evidence for an effect of habitat and presence of Barred Owls on fecundity, but in both cases the confidence intervals for the regression coefficients overlapped zero. The lack of a strong signal regarding the effects of habitat and Barred Owls on fecundity in the meta-analysis was not surprising considering the high variation among study areas regarding the importance of the habitat and the highly variable number of detections of Barred Owls among study areas (Appendix B). The meta-analysis also provided little evidence that ownership, climate, or weather had strong effects on fecundity.

We did not monitor prey abundance on all our study areas, but some lines of evidence sug-gest that the high fecundity of Spotted Owls on the east slope of the Cascades in Washington could be due to particularly high abundance or availability of preferred prey such as flying squirrels and woodrats (Lehmkuhl et al. 2006a, b). In addition, the understory shrub layer in forests on the east slope of the Cascades tends to be less dense than in forests in western Washington and Oregon, which may make it easier for Spotted Owls to capture prey in forests on the east slope. Tests of the prey abundance and availability hypotheses will likely prove difficult, but one obvious need is to initiate studies to better evaluate annual variation in the total biomass of prey available to Spotted Owls in different study areas.

We identified three major difficulties in the approach we used to model fecundity in the present analysis and previous meta-analyses. First, it was difficult to establish the effects of other variables in the presence of the strong even–odd year fluctuations in fecundity during the 1990s. If no adjustment is made for these even–odd year effects, the residual variation is large and negatively auto-correlated over time, which overwhelms the effects of any other covariate. In addition, because the even–odd year effect started to dissipate after about 2000, models that included the even–odd year effect had large residuals, which in turn made it difficult to detect the effects of other covariates.

Second, some of our covariates were highly correlated and in many cases also reflected time variation. For example, the BO covariate was negatively correlated with temporal trends because the proportion of territories on which Barred Owls were detected increased on most study areas over time (Appendix B). The habitat covariate was also somewhat correlated with time because it mainly reflected habitat loss over time.

Finally, some of the covariates we investigated were likely influential at the level of the individual territory, but in this analysis we modeled average effect across populations (study areas). For example, habitat and Barred Owls may have a strong effect on fecundity of individuals, but this could be masked by using yearly averages, particularly in conjunction with the strong annual variation in fecundity observed in our study. The above problems are likely to be present in any study of a species with a cyclic pattern of fecundity or with highly correlated covariates. There is no easy solution to these problems, except to recognize that they occur, and to avoid the inclusion of highly correlated covariates in the same models.

Apparent Survival

Annual recapture probabilities of territorial Spotted Owls in our study areas generally ranged from 0.70 to 0.90, within the range of estimates reported in previous studies of Spotted Owls (Burnham et al. 1996, Anthony et al. 2006). With the exception of one study area (OLY), our results indicated that male and female Spotted Owls had similar survival rates. Studies of Ural Owls (*Strix uralensis*; Saurola 2003) and Tawny Owls (*S. aluco*; Karell et al. 2009) also indicated no gender differences in survival of these species as well (but see Millon et al. 2009). Gender differences in survival of birds have been attributed to many factors, including sexual differences in dispersal (Croxall et al. 1990), plumage attributes (Møller and Szép 2002), territorial defense (Clobert et al. 1988), and feeding behavior (Clobert et al. 1988). Because male Spotted Owls play the dominant role in territorial defense and feeding of the young, we predicted that, if anything, they would have lower survival than females. The pattern on the OLY study area was opposite to this expected result, which supported the alternative hypothesis that egg production, incubation, brooding, and nest defense had higher costs on the survival and site fidelity of females than did territorial defense and foraging by the male.

Results from our study areas also indicated that apparent survival was influenced by a number of other factors including age, time, Barred Owls, reproduction, and weather, depending on the study area in question. The age-specific pattern that we observed (lower survival in young birds) is typical of many, if not

most, species of birds (Clobert et al. 1988; Newton 1989; Saurola 1987, 2003; Martin 1995; Karell et al. 2009). In long-lived, territorial birds like Spotted Owls, higher adult survival is probably attributable to the acquisition of a territory, foraging experience, and familiarity with the foraging area (Newton 1989, Martin 1995), but tests of these hypotheses have not been conducted.

Our estimates of survival were generally comparable to those reported by Burnham et al. (1996) and Anthony et al. (2006) except that the range of estimates for each age group in our study was slightly narrower than in the earlier studies. Our results were also comparable to those for adult California Spotted Owls (Blakesley et al. 2001, Seamans et al. 2001, Franklin et al. 2004) and adult Mexican Spotted Owls in Arizona (Seamans et al. 1999). Results from all three subspecies of Spotted Owls throughout their geographic range indicated that survival rates were high, with relatively low annual variability, while fecundity was highly variable from year to year. This life history strategy has been referred to as "bet hedging" (Stearns 1976, Franklin et al. 2000, Gaillard et al. 2000), where natural selection favors adult survival at the expense of producing fewer young during years with unfavorable conditions. Selection for high and comparatively stable adult survival is important because sensitivity analyses on population dynamics of Northern Spotted Owls (Noon and Biles 1990, Lande 1991) and California Spotted Owls (Blakesley et al. 2001) indicated that annual rates of population change were most influenced by changes in adult survival.

One disturbing finding in our analysis was that estimates of apparent survival were declining on 10 of the 11 study areas (CLE, RAI, OLY, COA, HJA, TYE, CAS, NWC, HUP, GDR, Fig. 5, Table 22). In addition, fecundity was declining in 5 of the 11 areas (Table 22). Declines in apparent survival of Northern Spotted Owls on some study areas have been reported previously (Burnham et al. 1996, Anthony et al. 2006), but, in contrast to those studies, our results indicated that recent declines were occurring across the entire range of the subspecies, including the

TABLE 22

Summary of trends in demographic parameters for Northern Spotted Owls from
11 study areas in Washington, Oregon, and California, 1985–2008.

Study area	No. of territorial owls in 2008[a]	Fecundity	Apparent survival (Model-averaged)	$\hat{\lambda}$	$\Delta\lambda$[b]
Washington					
CLE	18	Declining	Declining	0.937	Declining
RAI	36	Increasing	Declining	0.929	Declining
OLY	54	Stable	Declining	0.957	Declining
Oregon					
COA	105	Increasing	Declining since 1998	0.966	Declining
HJA	152	Increasing	Declining since 1997	0.977	Declining
TYE	123	Stable	Declining since 2000	0.996	Stationary
KLA	136	Declining	Stable	0.990	Stationary
CAS	83	Declining	Declining since 2000	0.982	Stationary
California					
NWC	84	Declining	Declining	0.983	Declining
HUP	51	Stable	Declining since 2004	0.989	Stationary
GDR	125	Declining	Declining	0.972	Declining

[a] Counts are based on banded territorial owls used in the analysis of $\hat{\lambda}$ and do not include owls that were not banded or whose bands were not confirmed.

[b] Population trends are based on estimates of realized population change (Δ_t).

southern portion. Estimated declines in adult survival were most precipitous in Washington, where annual apparent survival rates were <0.80 in recent years (Fig. 5A), a rate that may not allow for sustainable populations with current rates of fecundity and recruitment (Noon and Biles 1990, Lande 1991). In addition, the declines in adult survival and fecundity in Oregon have occurred predominantly within the last five years (Fig. 5B) and were not observed in the previous analysis of data from Oregon (Anthony et al. 2006). Compared to study areas farther north, declines in survival on the GDR and NWC study areas in California were more gradual and over a longer period of years. Collectively, the declines in apparent survival of Northern Spotted Owls across much of the subspecies' range are cause for concern because

Spotted Owl populations are most sensitive to changes in adult survival rates (Noon and Biles 1990, Lande 1991).

Anthony et al. (2006) found evidence of a negative Barred Owl effect on apparent survival of Spotted Owls in only 2 of the 14 study areas they examined. In our analysis of data from individual study areas, the percent of Spotted Owl territories with Barred Owl detections had a negative effect on apparent survival of Spotted Owls on 6 of 11 areas examined (RAI, OLY, COA, HJA, GDR, NWC), with a weak or negligible effect on the other five areas (CLE, TYE, KLA, CAS, HUP). Thus, our results suggest that the negative effect of Barred Owls on survival of Spotted Owls may be increasing as Barred Owls continue to invade and increase in numbers in our study areas (Appendix B).

In the meta-analysis of apparent survival, we found differences among study areas and ecoregions, and considerable annual variation in adult survival. Apparent survival rates were higher in the Oregon Cascades Douglas-fir, Oregon Coastal Douglas-fir, and California Coast ecoregions compared to the Mixed-conifer ecoregions in Washington and Oregon/California. The meta-analysis also provided evidence of a downward trend in survival for all study areas, which was expected given that our analyses of the individual study areas indicated declining survival rates on 10 of 11 areas. The overall decline in survival suggests a further deterioration of the situation reported by Anthony et al. (2006), who found that declines in survival were limited primarily to study areas in Washington.

The best random effects models in the meta-analysis suggested that reproduction in the previous year and the proportion of territories with Barred Owl detections both had negative effects on survival. We found some evidence that early nesting season precipitation had a negative effect on apparent survival but there was little to no evidence that the Pacific Decadal Oscillation, Southern Oscillation Index, nesting season temperature, percent cover of habitat, ownership, or latitude were associated with survival. It was not surprising that we did not find much evidence for an effect of weather in the meta-analysis because a previous analysis of demographic data and weather variables from six of our study areas indicated that the association of apparent survival with weather and climate covariates was quite variable among areas (Glenn 2009, Glenn et al. 2010, 2011). The lack of association between survival and most weather covariates suggests that Spotted Owls are able to cope physiologically with a fairly broad range of adverse weather conditions before their survival is affected. Romero et al. (2000) proposed a similar hypothesis regarding the effects of weather on reproduction of three species of Arctic passerines. If survival is affected only by the most extreme weather events, which occur at unpredictable times, detection of these effects will likely require hierarchical analyses to evaluate

the influence of within-year or within-season weather events (Rotenberry and Wiens 1991).

Annual Rate of Population Change and Realized Rates of Population Change

Individual Study Areas

Our estimates of λ were <1.0 for all study areas (range = 0.929 to 0.996), and there was strong evidence that populations declined on 7 of the 11 areas that we examined (RAI, OLY, CLE, COA, HJA, NWC, GDR). On the other four areas (TYE, KLA, CAS, HUP), either populations were stable or the precision of the estimates was not sufficient to detect declines. The number of territorial owls detected on all 11 areas was lower at the end of the study than at the beginning, and few territorial owls could be found on some of the study areas in 2008 (Table 22). Estimated rates of decline were highest for study areas in Washington (RAI, OLY, CLE) and the COA study area in Oregon. The weighted mean estimate of λ for all 11 study areas was 0.971, indicating an average population decline of 2.9% per year during the years 1990 to 2006. An average annual decline of 2.9% is lower than the 3.7% reported by Anthony et al. (2006), but the rates are not directly comparable because Anthony et al. (2006) examined a different series of years and because two of the study areas in their analysis were discontinued (WEN, WSR) and not included in our analysis. In our analysis, rates of population decline for individual study areas were slightly higher than those reported by Anthony et al., who found that populations on 9 of 13 study areas were declining. In California, Franklin et al. (2004) found that estimates of λ_{RJS} for California Spotted Owls were negative on four of five study areas examined, but in all five cases the 95% confidence intervals on λ overlapped 1.0. Franklin et al. (2004:33) concluded that either ". . . the populations were stationary or the estimates of λ_t were not sufficiently precise to detect declines if they occurred."

Our estimates of λ apply only to the years from which the data were analyzed, which spanned the 16-year period from 1990 to 2006 (Table 19). Any predictions about past or future trajectories of Spotted Owl populations on our study areas are risky. Also, the estimates of λ are mean estimates of the annual rate of population change in the number of territorial Spotted Owls on the study areas, and the estimates of λ_t for each study area varied considerably. Consequently, we attempted to illustrate how annual changes in λ_t influenced trends in population numbers by estimating realized population changes, Δ_t, for each study area. Based on these estimates, populations on the CLE, RAI, OLY, and COA study areas declined 40 to 60% during the last 15+ years, and populations on HJA, NWC, and GDR declined by 20 to 30%. Populations of territorial owls on the TYE, KLA, CAS, and HUP study areas declined 5 to 15%, but confidence intervals for these estimates substantially overlapped 1.0, and precision of the estimates was not sufficient to detect such small declines. Both the timing of the population declines and the rates of decline differed among study areas (Fig. 10). Thus, there was no evidence that population declines were synchronized among study areas, even though some of the study areas were relatively close together (e.g., COA, TYE, KLA), and marked individuals from one study area were occasionally re-sighted in another study area. The number of populations that declined and the rate of decline on study areas in Washington and northern Oregon were noteworthy and should be cause for concern for the long-term sustainability of Northern Spotted Owl populations throughout the range of the subspecies.

Meta-analysis of Annual Rate of Population Change

In the meta-analysis of λ, we found differences among ecoregions and a negative effect of Barred Owls on survival. Apparent survival was highest in the Oregon Coast Douglas-fir ecoregion, which was expected given that the Oregon Coast Range study area also had higher survival in the meta-analysis of survival. Apparent survival and λ were lowest in the Douglas-fir and Mixed-conifer ecoregion in Washington, and

recruitment was highest for the Oregon/California Mixed-conifer region. There was weak evidence that apparent survival was related to the percent cover of suitable owl habitat on four of eight study areas, but there was no evidence that weather or land ownership influenced apparent survival in the meta-analyses of λ. In contrast, there was evidence that the amount of suitable habitat within study areas had a positive influence on recruitment, and recruitment was highest for study areas on federally owned lands that had the highest proportions of suitable owl habitat. Positive associations between the percent cover of suitable owl habitat and survival and recruitment were expected because previous studies (Franklin et al. 2000, Olson et al. 2004, Dugger et al. 2005) have also found positive associations between apparent survival or fecundity and the amount of older forests surrounding Spotted Owl nest sites. However, given the importance of habitat in most previous studies of Spotted Owls, we were surprised that the percent cover of suitable habitat was not included in the top models for all study areas. Weak effects of habitat in our analysis could be the result of using habitat as a study area covariate as opposed to a site-specific covariate. The area-specific habitat covariate may have obscured relationships that could only be detected with finer-scale analyses of survival and fecundity at the scale of the owl home range.

In the meta-analysis of λ, we asked: Is temporal variation in λ_t determined primarily by variation in φ_t, f_t, or both? This general question is relevant to management because the answer may provide guidance regarding which population parameter(s) managers should focus on most when designing habitat management plans. In addition, there is some basis for prediction regarding the most important population parameters for species like Spotted Owls based on previous research on evolution of life history strategies in animals. In mammals and birds with long life spans, such as Spotted Owls, population dynamics are typically characterized by (1) rates of population change that are most sensitive to changes in adult survival, and (2) adult survival that exhibits a relatively small amount of temporal variation compared to temporal variation in recruitment (Pfister 1998; Gaillard et al. 1998, 2000; Gaillard and Yoccoz 2003). The degree to which annual variation in population change reflects variation in one parameter or another is a function of both the sensitivity of λ to that parameter and temporal variation in the parameter. Based on these patterns, we predicted there would be small temporal variability in adult survival compared to recruitment. The plots of year-specific estimates of λ_t, φ_t, and f_t provided illustrations of the temporal variation in annual population changes and its two primary components (φ_t and f_t; Fig. 12).

Although it was not our objective to draw inferences about whether survival or recruitment was more "important" to population change (see Hines and Nichols 2002 for discussion of this topic), we were interested in whether survival of territorial adults varied so little over time that most temporal variation in λ_t was produced by temporal variation in recruitment. This prediction did not hold true for Northern Spotted Owls because survival of adults varied considerably among years (range $\approx$ 0.70 to 0.90). Because of the importance of adult survival to annual population change (Lande 1988, Noon and Biles 1990), the observed variation in adult survival often corresponded closely to annual variation in λ and was most noticeable where populations were declining the most, especially study areas in Washington. However, the annual variation in apparent survival in our study was not nearly as great as annual variation in reproduction, so our results do fit the pattern usually observed in long-lived vertebrates, where survival is relatively constant compared to fecundity (Stearns 1976, Franklin et al. 2000, Gaillard et al. 2000).

Status of Owl Populations in the Eight NWFP Monitoring Areas

Eight of the study areas in our analysis (CLE, OLY, COA, HJA, TYE, KLA, CAS, NWC) are part of the effectiveness monitoring program for the Northern Spotted Owl in the Northwest Forest Plan (NWFP; Lint et al. 1999). As such, these areas are of special interest to the federal agencies charged with management of the owl. Our analysis indicated that populations on five of these study areas (CLE, OLY, COA, HJA, NWC) were declining during our study. Point estimates of λ on the other three areas (TYE, KLA, CAS) were <1.0, but the 95% confidence intervals on the estimates of λ broadly overlapped 1.0, so we could not reject the hypothesis that those populations were stationary. The weighted mean λ for the eight monitoring areas was 0.972 (SE = 0.006), which indicated that populations on those areas declined on average 2.8% per year during the 16-year study period.

Our results from the meta-analyses of fecundity and apparent survival were similar regardless of whether we used the entire sample of 11 study areas or limited the analysis to the eight NWFP monitoring areas. Therefore, we suggest that future analyses of the data from Northern Spotted Owl demography study areas be conducted only on the entire sample. Conducting a single analysis of all the data will greatly simplify the cooperative approach without losing any important information.

Associations Between Demographic Parameters and Covariates

Determination of cause–effect relationships is not possible with observational studies like ours. Rather, we attempted to assess the relative strength of associations between vital rates of owls and various environmental parameters such as habitat, weather, and presence of Barred Owls. It is implicit in this type of analysis that strong associations between vital rates and environmental factors are likely indicative of cause–effect relationships. Testing for associations is a common approach in ecology, where experimental tests of cause–effect relationships are difficult or impossible to conduct. Previous meta-analyses of demography of Northern Spotted Owls lacked the ability to assess potential processes responsible for causes of population declines. As a result, Anthony et al. (2006) recommended the development and use of biological covariates to help explain the variability in demographic rates and better understand the possible reasons for population changes. Consequently, we devoted considerable time to the development and refinement of covariates for evaluating the potential effects of reproduction, Barred Owls, climate, and percent cover of suitable owl habitat on fecundity, apparent survival, and recruitment at the population (study area) scale. Reproduction and Barred Owl covariates were previously investigated in the Anthony et al. (2006) analysis, but the climate and habitat covariates were new to our analysis. We also spent considerable time trying to develop a covariate for Barred Owls that was both time- and territory- or individual-specific, but inclusion of such a covariate proved infeasible in our analysis. Use of territory-specific covariates has proven feasible only in studies such as those conducted by Olson et al. (2004, 2005), Bailey et al. (2009), and Dugger et al. (2005), where the frame of reference is the individual territory as opposed to the study area or region. The area-specific Barred Owl covariate that we used differed from the covariate used by Anthony et al. (2006) in that our metric was based on Barred Owl detections anywhere within a 1-km radius of any of the historic activity centers in each Spotted Owl territory (see Methods for more details), as opposed to just the most recently occupied activity center. We used the new Barred Owl covariate because it may be a better indicator of the potential influence of Barred Owls on Spotted Owls in each territory.

Cost of Reproduction on Survival

There have been a number of correlative studies in which researchers found evidence that reproduction had negative effects on survival of breeding birds, including Western Gulls (*Larus occidentalis*; Pyle et al. 1997), Greater Flamingos (*Phoenicopterus ruber*; Tavecchia et al. 2001), Great Tits (*Parus major*; McCleery et al. 1996), and Lesser Scaup (*Aythya affinis*; Rotella et al. 2003). Anthony et al. (2006) found

that apparent survival of Northern Spotted Owls was negatively related to the mean number of young produced in the previous summer on some study areas in Washington and higher-elevation areas in Oregon. They hypothesized that negative correlations between survival and reproduction suggested a cost of reproduction, with the ultimate factor being weather-related. Although the reproduction covariate was not included in the top or competitive models for most individual study areas in our analysis, it was a factor in the best random effects model in the meta-analysis of survival. Based on this result, we concluded that there was evidence of a negative effect of reproduction on survival, even though the reproduction covariate did not explain a large amount of the annual variation in adult survival. The potential effect of reproduction on apparent survival did not appear to be related to the recent and widespread declines in Spotted Owl populations; however, it may be a contributing factor to some of the population declines, and this relationship needs further investigation. If a cost of reproduction is important in Spotted Owls, the proximate causes could include increased exposure to predation or increased energy expenditure while foraging, feeding young, and defending the territory. These factors have all been proposed as potential costs associated with reproduction in other birds (Newton 1989), but have been experimentally tested in only a few cases, with mixed results (Cichoń et al. 1998).

Weather and Climate

Several studies have documented associations between fecundity or apparent survival of Northern Spotted Owls and seasonal weather patterns (Wagner et al. 1996, Franklin et al. 2000, Olson et al. 2004, Glenn 2009, Glenn et al. 2010, 2011). Our results indicated that associations between fecundity, apparent survival, or recruitment and weather covariates varied among study areas. Fecundity was positively associated with mean temperature during the early nesting season on four of our study areas (RAI, COA, CAS, GDR). The positive association between fecundity and warm weather during the early nesting season has also been noted in several previous studies in which researchers used territory-based analyses to examine the effects of weather on fecundity of Spotted Owls(Wagner et al. 1996, Franklin et al. 2000, Olson et al. 2004, Glenn et al. *In press*). In addition, there was some evidence that fecundity was negatively associated with mean precipitation during the early nesting season on the KLA, CLE, and NWC study areas, and mean temperature during the late nesting season had a negative association with fecundity on TYE. Our results, and those of others (Franklin et al. 2000, Olson et al. 2004, Glenn et al. *In press*), suggest that years of high precipitation and low temperatures during the early nesting season can have a negative effect on fecundity of Northern Spotted Owls.

In our meta-analysis of survival, we detected a positive association between apparent survival and the Pacific Decadal Oscillation, and a negative association between apparent survival and early nesting season precipitation, but these associations were not strong. Similarly, the meta-analysis of λ indicated a positive association of apparent survival with the Pacific Decadal Oscillation, but no evidence for an association between recruitment and any of the climate covariates. (Glenn et al. 2010) reported a similar association between λ and the Pacific Decadal Oscillation on a subset of the study areas in our analysis. Positive values of the Pacific Decadal Oscillation are associated with lower than average rainfall and higher than average temperatures (Parson et al. 2001). We did not find evidence for any other associations between survival or recruitment of Northern Spotted Owls and weather or climate covariates in the meta-analyses. Lack of effects was not surprising because weather and climate varied considerably across the range of the Northern Spotted Owl, even within the same year (Glenn et al. 2010). Thus, analyses of potential associations between demographic rates and weather and climate

covariates on individual study areas may reveal patterns that were obscured in our meta-analysis of multiple study areas.

In summary, our analysis of climate covariates indicated the most evidence for a positive association between fecundity and mean temperature during the early nesting season, and a negative association between fecundity and mean precipitation during the early nesting season. We found little evidence for effects of weather on apparent survival and recruitment, and the only climate variable for which we found a positive association with apparent survival was the Pacific Decadal Oscillation. We concluded that weather and climate may contribute to lower demographic rates for some areas in some years, but the effects were not sufficient to explain the major population declines that have occurred during the last 15 to 20 years.

Barred Owls

The number of Barred Owl detections in our study areas has increased dramatically during the last two decades (Appendix B). The increase in Barred Owls has been most noticeable in Washington and Oregon, but has become apparent in northern California as well (Dark et al. 1998, Kelly 2001, Kelly et al. 2003). Invasion and rapid population growth of this congeneric species throughout the range of the Northern Spotted Owl has led to concerns of high potential for competition between the two species. Recent studies have also documented a negative association between occupancy of nesting territories (Kelly et al. 2003, Olson et al. 2005), fecundity (Olson et al. 2004), and apparent survival (Anthony et al. 2006) in some areas in relation to the presence of Barred Owls near nesting areas of Spotted Owls. Consequently, we hypothesized that demographic rates would be negatively associated with the presence of Barred Owls within 1 km of activity centers of Spotted Owls.

We found evidence that fecundity was negatively associated with the presence of Barred

Owls on the CAS, COA, KLA, and GDR study areas. Moreover, apparent survival was negatively associated with the presence of Barred Owls on the RAI, OLY, COA, HJA, GDR, and NWC study areas in both analyses of individual study areas and the meta-analysis. The meta-analysis of λ also indicated a negative association of apparent survival and recruitment with the proportion of territories with Barred Owl detections, but the evidence for a relationship with recruitment was weak. We also found evidence for a negative association of re-sighting probabilities of Spotted Owls when Barred Owls were detected near Spotted Owl nest areas on some of the individual study areas. In summary, we found evidence of negative relationships between demographic rates of Spotted Owls and the presence of Barred Owls on most study areas; therefore, our initial hypothesis was confirmed at least on some study areas. We suspect that the variable relationships between vital rates of Spotted Owls and the presence of Barred Owls were primarily due to the variable detection rates and arrival dates of Barred Owls invading the study areas (Appendix B). Another explanation for the inconsistent, and in some cases weak, associations between vital rates of Spotted Owls and detections of Barred Owls is that our BO covariate was coarse in scale (year-specific only) and was applied at the population scale and not the individual territory scale. Consequently, we believe the influence of Barred Owls on demography of Spotted Owls is likely stronger than was indicated by our analyses. There is a need to develop a covariate for Barred Owls that is both year- and territory-specific (Anthony et al. 2006). Our results support the findings of previous studies that have also reported evidence for negative associations of demographic performance of Spotted Owls when Barred Owls were detected near their nest areas (Kelly et al. 2003; Olson et al. 2004, 2005; Anthony et al. 2006). In addition, Olson et al. (2005) found evidence that occupancy and colonization rates of Spotted Owl territories were negatively associated with detections of

Barred Owls. In another territory-specific study, K. Dugger et al. (In press) found evidence that extinction rates of Spotted Owl territories were higher on territories with Barred Owl detections, and this effect was stronger as the amount of habitat decreased. The latter results suggested an additive effect of decreasing habitat and presence of Barred Owls on demographic performance of Spotted Owls.

Taken together, results of our current study and previous studies do not prove a causal effect of Barred Owls on the demography of Northern Spotted Owls. However, the consistency of the negative associations between Spotted Owl demographic rates and presence of Barred Owls in multiple studies lends support to the conclusion that Barred Owls are having a negative effect on spotted owl populations. Of the various factors we investigated to ascertain potential effects on demographic rates of Northern Spotted Owls, the mostly negative associations with the presence of Barred Owls were the strongest and most consistent factor among study areas. The negative associations with Barred Owls were more numerous and stronger in our analysis than those reported by Anthony et al. (2006), and corresponded with the increase in detections of Barred Owls in the last five years on our study areas. The increasing evidence for a Barred Owl effect suggests that recent declines in fecundity, apparent survival, and populations of Spotted Owls on our study areas are at least partly due to interactions with Barred Owls. However, we cannot rule out the potential influence of continued declines in habitat as another factor contributing to population declines (see below).

Habitat

Our investigation of the potential influence of habitat on demographic rates of Northern Spotted Owls was both challenging and problematic for a number of reasons. First, comparable vegetation maps from satellite imagery for the entire range of the subspecies were not available, and it was clear during the workshop that the imagery for California was developed with different criteria and was different from the vegetation map of Washington and Oregon. As a result, we excluded the California study areas in the meta-analysis of demographic rates with the habitat covariate. Second, the available map for Oregon and Washington did not span the entire length of time that the demographic studies were conducted, so we had to estimate the amount of suitable owl habitat that was present on the study areas both prior to and after 1996, when the best map was available. We estimated the amount of habitat that was lost due to harvest and wildfires during the time of the studies with a change detection algorithm (see Methods section). Third, there may have been some small amount of forest that became suitable owl habitat as a result of forest re-growth during our studies, but we could not readily identify these forests to be able to adjust our estimates accordingly. Fourth, the maps that we used characterized forest vegetation at landscape scales and did not characterize the understory structure, which has been shown to be important for Spotted Owls and their primary prey (Carey et al. 1992, Rosenberg and Anthony 1992, Buchanan et al. 1995, LaHaye and Gutiérrez 1999, Lehmkuhl et al. 2006b).

While the amount of suitable habitat on some study areas in Oregon had a positive effect on reproduction, there was little evidence for a consistent effect of habitat on fecundity for all areas in Washington and Oregon from the meta-analysis. The absence of a strong association between the amount of habitat and fecundity was not entirely surprising considering that two previous studies found evidence that "habitat fitness" for Spotted Owls increased in landscape configurations that included a mixture of old forests and edge (Franklin et al. 2000, Olson et al. 2005, but see Dugger et al. 2005). Whether inclusion of a forest edge covariate in our analysis would have made a difference in the outcome is unclear, but

inclusion of such a covariate should be considered in future analyses.

In the meta-analysis of survival, apparent survival was positively related to the percent cover of suitable owl habitat within the study area boundaries, but the 95% confidence intervals overlapped zero, indicating that the evidence for an association was weak. The habitat covariate was not included in the analysis of survival rates for individual study areas, which was an oversight during the development of the protocol (see below). Such analyses should be considered in the next major analysis of demographic data from Spotted Owls. In the meta-analysis of λ, apparent survival was related positively to the percent cover of suitable habitat in the CLE, COA, HJA, and TYE study areas, as 95% confidence intervals for the regression coefficients for the habitat covariate barely overlapped zero. More importantly, we found a positive relationship between recruitment and the percent cover of suitable owl habitat within the study areas in the meta-analysis of λ. Recruitment was also highest on federally owned lands where the amount of suitable habitat was highest (Davis and Lint 2005). One possible explanation for the latter result is that more habitat within the study areas provided areas where non-territorial owls could occupy and survive until they were able to recruit into the territorial population.

A number of territory-specific studies of Spotted Owls have reported fairly strong associations between the amount of suitable habitat and demographic rates of Spotted Owls. The fact that we found relatively weak associations between the amount of habitat and demographic rates suggests that our area-specific covariate was too coarse to reveal actual relationships that were acting at the scale of the individual owl territory. Our conclusion should not be used to infer that the amount of old forest (suitable owl habitat) is not important to the demography of the Spotted Owl, because other studies have documented positive associations between demography and the amount of old forest surrounding nest sites of Spotted

Owls. For example, apparent survival was positively related to the amount of old forest surrounding nest sites in territory-specific studies of Spotted Owls in northwestern California (Franklin et al. 2000) and southern Oregon (Dugger et al. 2005), In the territory-specific studies conducted by Franklin et al. (2000) and Olson et al. (2004), large areas of mature and old forest interspersed with openings provided the best habitat for Northern Spotted Owls in northwestern California and the Oregon Coast Ranges. In southern Oregon, Dugger et al. (2005) found that reproductive rates of Spotted Owls were positively related to the proportion of old-growth forest within a 730-m-radius circle around nest sites. In the Sierra Nevada of California, Seamans and Gutiérrez (2007) observed higher colonization and lower extinction rates for California Spotted Owls on territories with more mature conifer forest. In the above studies, analyses were conducted at the scale of owl territories within study areas and with a smaller scale of habitat mapping from aerial photographs; the results of those studies were more definitive than our study, which was at the scale of entire study areas (populations). Also, recent analyses of occupancy dynamics of Northern Spotted Owls in the southern Cascades of Oregon indicated that there was an additive and negative effect of Barred Owls and decreased amounts of habitat on occupancy and colonization, and a positive effect on extinction of nesting territories (Dugger et al. In press). The latter results suggest that it may be necessary to conserve even more old forest habitat than is currently protected, if the objective is to increase the likelihood that Spotted Owls will be able to persist in the face of potential competition with Barred Owls for space, habitat, or prey. Competition theory predicts that more habitat is necessary if two species are to persist when they are in direct competition (Levins and Culver 1971, Horn and MacArthur 1972), an important consideration in the conservation of Northern Spotted Owls. Carrete et al. (2005) recommended an increase in suitable habitat for two potentially competing raptors, the Golden Eagle (*Aquila chrysaetos*) and Bonelli's Eagle (*A. fasciata*) in southern Spain. Last, it is well documented that Northern Spotted Owls select older forests for nesting (Hershey et al. 1998, Swindle et al. 1999), and roosting and foraging (Forsman et al. 1984, Thomas et al. 1990, Bart and Forsman 1992, Herter et al. 2002, Glenn et al. 2004, Forsman et al. 2005) throughout most of their range, so these forests are important to their survival and population persistence. Selection for the oldest available forest is consistent even within managed forests on private lands in northwestern California, where Diller and Thome (1999) and Thome et al. (2000) found that Spotted Owls usually occurred in the oldest available forests. Researchers studying California Spotted Owls have also reported strong associations with older forests for nesting, roosting, and foraging (LaHaye et al. 1997, LaHaye and Gutiérrez 1999). Consequently, despite the weak associations between demographic rates and habitat in our analysis, it would be incorrect to conclude from our results that old forest vegetation is not important to Northern Spotted Owls.

Potential Biases in Estimates of Demographic Parameters

Numerous authors have discussed possible biases associated with estimates of fecundity or survival from long-term demography studies of Northern Spotted Owls (Raphael et al. 1996, Van Deusen et al. 1998, Manly et al. 1999, Boyce et al. 2005, Loehle et al. 2005). In some cases, these critiques resulted in rigorous rebuttals (Franklin et al. 2006). Because parameter bias could have important effects on development of effective conservation and management strategies, we discuss potential sources of bias in our estimates of fecundity and apparent survival below.

Fecundity

Estimates of fecundity can be biased if territorial females are present on the study area but are not detected in any given year. If the undetected territorial females nest successfully, fecundity could be underestimated. If undetected birds do not nest, or nest and fail, fecundity is overestimated. These two sources of bias may cancel each other out because both scenarios can happen in the same year, but we suspect that the positive bias is slightly more prevalent than the negative bias because non-nesting females and females that nest and fail tend to be more difficult to detect than nesting females. However, re-sighting probabilities of owls in our study were typically >0.75, so the frequency of missing data on reproduction in most years was small. Even if there was a bias in our estimates of fecundity, this bias should have been consistent among years and study areas. Therefore, any small positive or negative bias in our estimates of fecundity should not have confounded any analyses in which we examined the effects of time, age, study area, geographic region, latitude, Barred Owls, climate, or habitat on fecundity.

Apparent Survival

Temporary or permanent emigration, heterogeneity in recapture probabilities, and band loss are the primary factors that may create biases or lack of precision in estimates of apparent survival from analysis of capture-recapture data. Two of these potential biases were investigated by Manly et al. (1999), who used computer simulations with data from Northern Spotted Owls in the eastern Cascades of Washington. Variation in recapture probabilities for nesting and non-nesting owls, temporary emigration, and dependent captures of both members of a breeding pair had little effect on estimates of apparent survival, although temporary emigration can cause lower apparent survival estimates for the last few years of a study. In addition, the combination of high recapture and survival probabilities in our study likely reduced any bias associated with heterogeneity of recapture probabilities (Pollock et al. 1990, Hwang and

Chao 1995). As for permanent emigration, Forsman et al. (2002) studied dispersal of territorial Spotted Owls on a subset of our study areas and estimated that only about 6.6% of resident owls dispersed from their territories each year, and most of those individuals were relocated on adjacent territories within the boundaries of our survey areas. Nevertheless, there were undoubtedly some individuals that dispersed and went undetected at the edges of our study areas, and to this extent, our estimates of apparent survival may have been biased low as an index of true survival.

Annual Rate of Population Change

Our use of the reparameterized Jolly–Seber method (RJS; Pradel 1996) to estimate the annual finite rate of population change (λ_{RJS}) was a departure from earlier analyses of Spotted Owls, in which researchers used Leslie projection matrices (PM; Caswell 2001) to estimate λ_{PM} (Anderson and Burnham 1992; LaHaye et al. 1992; Burnham et al. 1996; Seamans et al. 1999, 2002; Blakesley et al. 2001). Estimates of λ_{PM} were thought to be biased low in these studies because of permanent emigration of juveniles from study areas (Raphael et al. 1996, Boyce et al. 2005). In contrast, the Pradel (1996) method of estimating λ_{RJS} uses survival estimates from territorial owls only, so it is subject to less bias than the Leslie projection matrix models (λ_{PM}) for use in capture–recapture studies of Spotted Owls (Hines and Nichols 2002, Franklin et al. 2004, Anthony et al. 2006). Estimation of λ_{RJS} assumes that study area boundaries are fixed throughout the study and that surveys of territorial owls are conducted on the same areas with similar effort each year. In other words, new owls are not recruited into, or previously sampled owls are not lost from the sample because of changes in survey area or methods. We used established protocols for surveying and identifying marked Spotted Owls (Franklin et al. 1996, Lint et al. 1999) to ensure that study areas were surveyed with approximately equal effort each year. In addition, the study areas had fixed geographical boundaries for inclusion of data from individual owls, and any expansion or contraction of study areas (Appendix A) was corrected for by modeling in program MARK (see Methods section). Thus, the primary assumptions for estimating λ_{RJS} from capture–recapture data from Spotted Owls were met. The Pradel method for estimating λ accounts for movement into and out of the study area and is less subject to bias caused by permanent emigration of marked owls, which is why the Pradel models may improve on the Leslie matrix model for estimating the annual rate of population change for Spotted Owls. If movements in and out of the study area are truly asymmetric, then the Pradel method should produce a high or low λ to reflect this (it is not a bias, but an accurate reflection of reality).

Last, band loss in our studies was near zero. Franklin et al. (1996) examined records from over 6,000 Northern Spotted Owls double-banded with a colored band and a numbered metal band, and found only two cases where colored bands were lost and no cases where the numbered metal band was lost. Based on the above assessments, we believe that any biases in our estimates of λ were small.

Estimating Goodness-of-Fit and Overdispersion

There are potential biases in the estimation of overdispersion (c) when the estimate is based on the global goodness-of-fit statistic from program RELEASE. The overall goodness-of-fit chi-square (χ^2) is comprised of three additive components: identifiable outliers, structural lack-of-fit, and lack of independence in capture histories (overdispersion). These three potential components of lack-of-fit have differing effects on bias and precision of parameter estimates.

Outliers and structural lack-of-fit can result in biased estimators of φ and λ_{RJS}, but do not result in inflated variances of these estimators. Moreover, these components of lack-of-fit do not result in, and hence are not part of, overdispersion. In contrast, overdispersion does not cause bias in the estimates of φ, p, or λ_{RJS}, but it does result in estimated sampling variances that are too small. Thus, one needs an estimate of overdispersion (c) to adjust (inflate) the estimated theoretical sampling variances and adjust model selection to QAIC$_c$. Estimates of overdispersion and the variance inflation factor from program RELEASE in previous analyses of capture–recapture data from Spotted Owls were biased high (e.g., Franklin et al. 2004, Anthony et al. 2006). As a result, sampling standard errors from those analyses were conservative in assessing the status of populations from the estimation of λ_{RJS} and corresponding 95% confidence intervals. We corrected for this overestimation of overdispersion in our analysis by using the median-$\hat{c}$ routine in program MARK to estimate overdispersion in addition to using program RELEASE to estimate overall goodness-of-fit. Estimates from the median-$\hat{c}$ routine of program MARK in our analyses ranged from $\hat{c} = 0.97$ to 1.17 compared to the range of estimates for overall goodness-of-fit (χ^2/df) from program RELEASE ($\hat{c} = 0.86$ to 3.02). Our results indicated that there was little overdispersion (lack of independence) in our capture–recapture data sets, and any overall lack-of-fit was due to outliers caused by temporary emigration and perhaps some structural lack-of-fit. Consequently, inflation of our estimates of SE(φ) and SE(λ) was minimal, and the true precision of our estimates was higher than those in previous analyses given equal sample sizes (Franklin et al. 2004, Anthony et al. 2006). Use of the median-$\hat{c}$ routine in program MARK to estimate overdispersion in our analyses was an important improvement over previous analyses. Estimates of goodness-of-fit from program RELEASE also indicated that our data fit the Cormack–Jolly–Seber open population model well, so we did not expect unacceptable biases due to lack-of-fit of the data to the model.

The covariates that we used to assess the effects of Barred Owls, habitat, weather, and climate on demographic parameters of Spotted

Owls were all study-area–specific variables, and in some cases they were not measured with the same degree of accuracy on all study areas. Use of area-specific covariates could explain why we sometimes found inconsistent or counterintuitive relationships between the covariates and demographic performance of Spotted Owls. Variable effort was a problem with the Barred Owl covariate because the amount of nocturnal survey varied among years and study areas, depending on whether it was a good nesting year for Spotted Owls. Surveyors sometimes did less night calling for Spotted Owls in good nesting years because many pairs of nesting Spotted Owls were easy to find by simply walking into their traditional nest areas and calling during the day. Variation in the amount of nocturnal calling surveys probably introduced methodological variation into the Barred Owl covariate, and lack of a species-specific survey for Barred Owls undoubtedly caused an underestimate of the number of Barred Owls present in all years. A recent study in which observers conducted a species-specific survey of Barred Owls in a Spotted Owl study area resulted in a ≈40% increase in the estimated number of territorial Barred Owls (Wiens et at. In press). An obvious solution to our problems with the Barred Owl covariate is to do a better job of measuring and standardizing all covariates in the future. For Barred Owls, improved procedures would require initiating species-specific surveys in which Barred Owl surveys are conducted independently of Spotted Owl surveys.

SUMMARY, CONCLUSIONS, AND RECOMMENDATIONS

The primary objectives of our investigation were to determine if survival rates and populations of Northern Spotted Owls were still declining, assess the influence of biological and meteorological covariates on demographic rates at the population scale, and provide estimates of recruitment rates. Our analyses indicated that fecundity and populations of Northern Spotted Owls have continued to decline in most parts of the range of the subspecies. Estimates of the annual rate of population change were <1.0 for all 11 study areas. Our finding that apparent survival rates were declining on 10 of the 11 study areas was of special concern because Spotted Owl populations are most sensitive to changes in adult survival (Noon and Biles 1990). We had some success in relating demographic rates to reproduction, weather, habitat, or Barred Owls on some study areas. In the analysis of fecundity, however, the amount of temporal variation explained by any one of these covariates was small due to the large temporal variation in fecundity. Temporal variation was not as problematic in the analyses of apparent survival and λ, because these parameters had much less temporal variation than fecundity. For the first time, we provided estimates of recruitment rates into the territorial population, which indicated that low recruitment in conjunction with low survival resulted in population declines. We also found a negative relationship between recruitment rates and the presence of Barred Owls and a positive relationship between recruitment and the amount of suitable owl habitat in the study areas. Recruitment was higher on federal lands where the amount of suitable owl habitat was generally highest. We concluded that there were several factors that contributed to declines in demographic rates of Northern Spotted Owls in any given year on any particular study area, and that these factors were spatially and temporally variable. Of these factors, the presence of Barred Owls appeared to be the strongest and most consistent factor. However, the reproduction covariate, weather/climate covariates, and percent cover of suitable habitat were also associated with demographic parameters on some study areas. Declining rates of apparent survival were the most likely proximate cause of population declines, but the ultimate factor(s) responsible for the declines in survival remained unclear and warrant further investigation. In addition, recruitment of new owls into the populations was often low on some study areas in some years and contributed to population declines. Future analyses should investigate the factors that affect survival of juvenile owls and their recruitment into the territorial population. All of these demographic parameters and the covariates that may affect them interact in a complex way in influencing annual rates of population change of Northern Spotted Owls. Our overall assessment is that reproduction and recruitment have not been sufficient to balance losses due to mortality and emigration, so many of the populations on our study areas have declined over the last two decades. The continuing decline of the Northern Spotted Owl on federal lands could be at least partly due to lag effects from the extensive harvest of old forest that occurred prior to 1990. However, the lag-effect hypothesis was not supported by ongoing declines among owl populations in national parks, where there was no habitat loss due to harvest at any time in the years before or during our study. Thus, we do not think the lag-effect hypothesis has much explanatory power for the continuing declines of Northern Spotted Owls.

Although the pattern was not consistent in all areas, there was strong evidence for a negative effect of Barred Owls on fecundity or survival of Spotted Owls in many of our study areas. This result was even more significant given that the actual effect of Barred Owls on fecundity of Spotted Owls was underestimated by our data. While our observational results do not demonstrate cause–effect relationships, they provide support for the hypothesis that the invasion of the range of the Spotted Owl by Barred Owls is at least partly the cause for the continued decline of Spotted Owls on federal lands. Our results also suggest that Barred Owl encroachment into western forests may make it difficult to insure the continued persistence of Northern Spotted Owls (see also Olson et al. 2004). The fact that Barred Owls are increasing and becoming an escalating threat to the persistence of Spotted Owls does not diminish the importance of habitat conservation for Spotted Owls and their prey. In fact, the existence of a new and potential competitor like the Barred Owl makes the protection of habitat even more important, since any loss of habitat will likely increase competitive pressure and result in further reductions in Spotted Owl populations (Horn and MacArthur 1972, Olson et al. 2004, Carrete et al. 2005). Manipulative experiments could provide future insights, and some authorities have suggested that removal experiments should be conducted on one or more study areas to better document the potential effects of competition between Barred and Spotted Owls (Courtney et al. 2004, Buchanan et al. 2007, Johnson et al. 2008). If conducted, manipulative experiments will almost certainly shed new light on relationships between Barred Owls and Spotted Owls.

The fact that the amount of spatial and process variation explained by all of the covariates in our analysis was small should not be interpreted to mean that habitat and climate are not important for Spotted Owls. To the contrary, several lines of evidence in our study and in studies conducted by others (Franklin et al. 2000, Olson et al. 2004, Dugger et al. 2005) show that habitat does influence demographic rates of Northern Spotted Owls. However, the poor performance of fixed effects models, which model temporal variation solely as a function of temporal covariates, should be discouraged in future analyses and replaced with improved random effects models that incorporate both environmental covariate(s) and temporal variation. In addition, we suggest that researchers need to consider the use of other covariates in future analyses. For example, there is considerable evidence that vital rates and population size of northern owls are strongly influenced by prey abundance (Korpimäki 1992, Rohner 1996, Hakkarainen et al. 1997). Unfortunately, we did not have long-term data on annual variation in prey abundance on any of our study areas, so we could not address the possible influence of trophic dynamics on owl demographic rates. We suggest, therefore, that studies of annual variation in numbers of small mammals be implemented on one or more of the demographic study areas in the future, so that the possible influence of prey abundance on owl demographic rates can be evaluated.

So, what can we glean from our results that can be translated into management recommendations? Our results and those of others referenced above consistently identify loss of habitat and Barred Owls as important stressors on populations of Northern Spotted Owls. In view of the continued decline of Spotted Owls in most study areas, it would be wise to preserve as much high quality habitat in late-successional forests for Spotted Owls as possible, distributed over as large an area as possible. This recommendation is comparable to one of the recovery goals in the final recovery plan for the Northern Spotted Owl (USDI Fish and Wildlife Service 2008), but we believe that a more inclusive definition of high-quality habitat is needed than the rather vague definition provided in the 2008 recovery plan. Much of the habitat occupied by Northern Spotted Owls and their prey does not fit the classical definition of "old-growth" as defined by Franklin and Spies (1991), and a narrow definition of habitat based on the Franklin and Spies criteria would exclude many areas currently occupied by Northern Spotted Owls. Second, we believe more information on competitive interactions between Spotted Owls and Barred Owls is needed. A recent study by D. Wiens at Oregon State University (pers. comm.) will provide some of this information for western Oregon, but similar information is needed for other parts of the range of the Spotted Owl. In addition, we support experimental removal of Barred Owls on at least one study area as a research project to test the hypothesis that competition is occurring between the two species. In theory, a Barred Owl removal experiment should result in competitive release of Spotted Owls, with subsequent increases in vital rates and density. Experimental removal of Barred Owls as part of a research program would also address one of the main recovery goals in the final recovery plan for Northern Spotted Owls (USDI Fish and Wildlife Service 2008). Finally, it is important that monitoring of Northern Spotted Owls be continued on study areas throughout the range of the subspecies, so that population status can be assessed periodically for the purposes of recovery planning and monitoring the effectiveness of the Northwest Forest Plan.

Appendices

APPENDIX A

Study areas included in the January 2009 analysis of demographic trends of Northern Spotted Owls.

Study area	Start year[a]	λ Start year	Expansion year[b]	Landowner[c]	Ecoregion	Latitude (°N)
Washington						
CLE	1989	1992	none	Mixed	Washington Mixed-conifer	46.996
RAI	1992	1993	1998	Mixed	Washington Douglas-fir	47.195
OLY	1990	1990	1994	Federal	Washington Douglas-fir	47.800
Oregon						
COA	1990	1992	none	Mixed	Oregon Coastal Douglas-fir	44.381
HJA	1988	1990	2000	Federal	Oregon Cascades Douglas-fir	44.213
TYE	1990	1990	none	Mixed	Oregon Coastal Douglas-fir	43.468
KLA	1990	1990	1998	Mixed	Oregon/California Mixed-conifer	42.736
CAS	1991	1992	2001	Federal	Oregon Cascades Douglas-fir	42.695
California						
NWC	1985	1988	none	Federal	Oregon/California Mixed-conifer	40.848
HUP	1992	1992	none	Tribal	Oregon/California Mixed-conifer	41.051
GDR	1990	1990	1998	Private	California Coast	41.122

[a] The Start year column indicates the first year in which we calculated estimates of fecundity and survival. The λ Start year column indicates the first year in which we calculated estimates of λ.

[b] Indicates year that study area was expanded, if any.

[c] Mixed = a mixture of Federal and private or state lands

APPENDIX B

Annual proportion of Spotted Owl territories with Barred Owls detections (BO covariate)
on study areas in Washington, Oregon, and California.

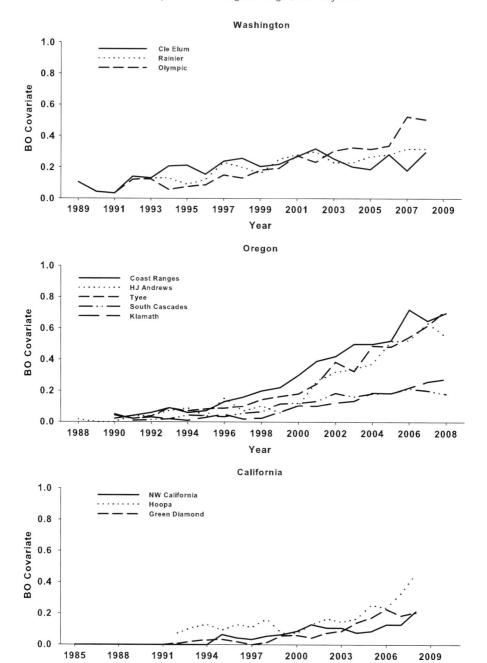

APPENDIX C

Habitat covariates used in analyses of Northern Spotted Owl vital rates and population growth rates.

Graph A illustrates the percent cover of suitable Spotted Owl habitat within 2.4 km of the annual activity centers of Spotted Owls used in meta-analyses of fecundity and survival (covariate HAB1). Graph B illustrates the percent cover of suitable Spotted Owl habitat within 2.4 km of the annual activity centers of Spotted Owls that were included in the meta-analysis of λ (HAB2). Graph C illustrates the percent cover of suitable Spotted Owl habitat within a 23-km radius of the annual activity centers of Spotted Owls that were included in the meta-analysis of λ, minus the area in HAB2 (HAB3). Abrupt changes in some lines represent one-time study area expansions or reductions included in the meta-analysis of λ.

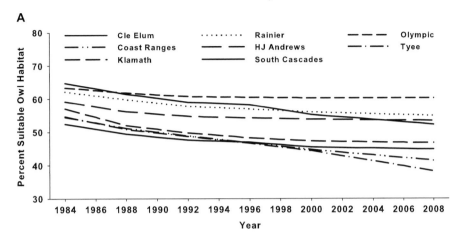

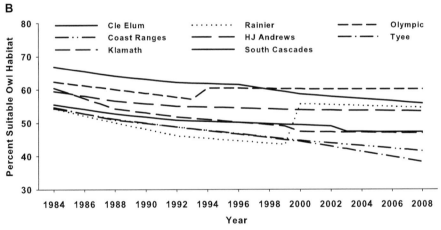

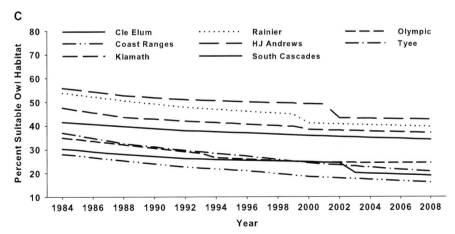

APPENDIX D

Reproductive covariate (number of young fledged/pair/yr) used to model survival, and recapture probabilities of Northern Spotted Owls on 11 study areas in Washington, Oregon, and California.

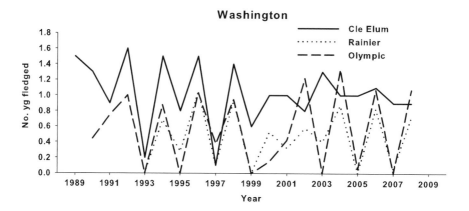

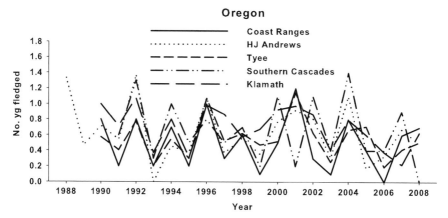

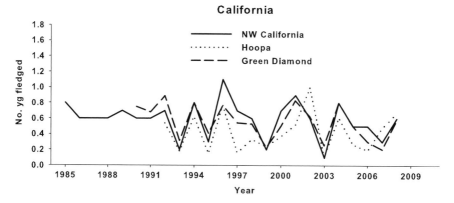

APPENDIX E

A priori models used in analysis of recapture probabilities (p) of Northern Spotted Owls on 11 demographic study areas in Washington, Oregon, and California.

Model[a]	Description of p structure
$p(A + s^*t)$	Additive age, sex, and time effects with interactions between sex and time
$p(.)$	Constant model (no effects)
$p(s)$	Sex effect
$p(R)$	Effect of annual reproduction in year t on p in year t
$p(R + s)$	Additive reproduction and sex effects
$p(t)$	Annual time effect
$p(s + t)$	Additive sex and time effects
$p(T)$	Linear time trend effect
$p(s + T)$	Additive sex and linear time trend effects
$p(BO)$	Barred Owl effect
$p(s + BO)$	Additive sex and Barred Owl effects
$p(R + s + BO)$	Additive sex, Barred Owl, and reproduction effects
$p(choice)$	Biologist's choice

[a] Model notation indicates structure for effects of age (A), sex (s), reproduction (R), time (t), linear time (T), percent of Spotted Owl territories with Barred Owl detections (BO), and biologist's choice (choice). Biologist's choice models included study-area–specific effects such as changes in methodology or subdivisions of study areas based on forest type or ease of access. Additive and interactive effects are indicated by a + sign or asterisk, respectively.

*A priori models used for analysis of apparent survival (φ) of Northern Spotted Owls on
11 demographic study areas in Washington, Oregon, and California.*

Analyses used the best *p* structure from the initial analysis for each area.

Model	Description of φ structure
$\varphi(.)$	Constant survival, no age, sex, or time effects
$\varphi[(S1 = S2 = A) + s]$	Sex effect only
$\varphi(S1, S2 = A)$	Age effect (S2 = A, S1 different)
$\varphi[(S1, S2 = A) + s]$	Age effect (S2 = A, S1 different), additive sex effect
$\varphi(S1 = S2, A)$	Age effect (S1 = S2, A different)
$\varphi[(S1 = S2, A) + s]$	Age effect (S1 = S2, A different), additive sex effect
$\varphi(S1, S2, A)$	Age effect (all classes different)
$\varphi[(S1, S2, A) + s]$	Age effect (all classes different), additive sex effect
$\varphi[(\text{models } 1–8) + t]$	Models from 1–8 above with additive time effect (t)
$\varphi[(\text{models } 1–8) + T]$	Models from 1–8 above with additive linear time trend (T)
$\varphi[(\text{models } 1–8) + TT]$	Models from 1–8 above with additive quadratic time trend (TT)
$\varphi[(\text{models } 1–8) + R]$	Models from 1–8 above with additive effect of reproduction in year t on survival in year $t + 1$ (R)
$\varphi[(\text{models } 1–8) + BO]$	Models from 1–8 above with Barred Owl effect (BO)
$\varphi[(\text{models } 1–8) + \text{change-point}]$	Models from 1–8 above with change-point at 2002 (CP)[a]
$\varphi[(\text{models } 1–8) + \text{cubic spline}]$	Models from 1–8 above with cubic spline (spline)[b]

[a] Change-point in 2004 using best model structure of (.), (T), or (TT).

[b] Cubic spline with knot midway between start year and 2002 and second knot at 2002.

APPENDIX G

A priori models used for meta-analysis of apparent survival (φ) and recapture probabilities (p) of adult Northern Spotted Owls on 11 demographic study areas in Washington, Oregon, and California.

Area effects (g) refer to study areas.

Model	Description of Model Structure
Global model	
1. $\varphi(g*t*s)\ p(g*t*s)$	Area, time, and sex with all interactions (global model)
Recapture	
2. $\varphi(g*t + s)\ p(g + t)$	φ(Area, time, and sex with area and time interactions) p(additive area and time)
3. $\varphi(g*t + s)\ p(R)^a$	φ(Area, time, and sex with area and time interactions) p(reproduction)
4. $\varphi(g*t + s)\ p(g + s + t)$	φ(Area, time, and sex with area and time interactions) p(additive area, time, and sex)
5. $\varphi(g*t + s)\ p(R + s)$	φ(Area, time, and sex with area and time interactions) p(additive reproduction and sex)
6. $\varphi(g*t + s)\ p[(g + t)*s]$	φ(Area, time, and sex with area and time interactions) p(additive area and time with different sex effects)
7. $\varphi(g*t + s)\ p(R*s)$	φ(Area, time, and sex with area and time interactions) p(interactive reproduction and sex)
8. $\varphi(g*t + s)\ p(BO)$	φ(Area, time, and sex with area and time interactions) p(BO)
9. $\varphi(g*t + s)\ p(BO + g)$	φ(Area, time, and sex with area and time interactions) p(BO + area)
Survival	
10. $\varphi(g + s)\ p(\text{best})$	φ(additive area and sex) p(best structure from 2–9 above)
11. $\varphi(g + s + t)\ p(\text{best})$	φ(additive area and sex and time) p(best structure from 2–9 above)
12. $\varphi(g*T + s)\ p(\text{best})$	φ(interactive area and linear time trend with additive sex effect) p(best structure from 2–9 above)
13. $\varphi(g + s + T)\ p(\text{best})$	φ(additive area, sex, and linear time trend) p(best structure from 2–9 above)
14. $\varphi(g*TT + s)\ p(\text{best})$	φ(interactive area and quadratic time trend with additive sex effect) p(best structure from 2–9 above)
15. $\varphi(g + TT + s)\ p(\text{best})$	φ(additive area, quadratic time trend, and sex effect) p(best structure from 2–9 above)
16. $\varphi(s + t)\ p(\text{best})$	φ(additive sex and time effects) p(best structure from 2–9 above)
17. $\varphi(s + T)\ p(\text{best})$	φ(additive sex and linear time trend effects) p(best structure from 2–9 above)
18. $\varphi(s + TT)\ p(\text{best})$	φ(additive sex and quadratic time trend effects) p(best structure from 2–9 above)
19. $\varphi(s)\ p(\text{best})$	φ(sex) p(best structure from 2–9 above)
20. $\varphi(s + BO)\ p(\text{best})$	φ(additive sex and BO effects) p(best structure from 2–9 above)
21. $\varphi(s + BO + g)\ p(\text{best})$	φ(additive sex, BO effects, and area) p(best structure from 2–9 above)

APPENDIX G (*continued*)

Model	Description of Model Structure
22. φ(s + BO*g) *p*(best)	φ(interactive BO effects and area effects with additive sex effect) *p*(best structure from 2–9 above)
23. φ(s + R) *p*(best)	φ(additive sex and reproduction effects) *p*(best structure from 2–9 above)
24. φ(s + R + g) *p*(best)	φ(additive sex, reproduction, and area effects) *p*(best structure from 2–9 above)
25. φ(s + R*g) *p*(best)	φ(interactive reproduction and area effects with additive sex effect) *p*(best structure from 2–9 above)
26. φ(s + BO + R) *p*(best)	φ(additive sex, reproduction, and BO effects) *p*(best structure from 2–9 above)
27. φ(s + BO + g + R) *p*(best)	φ(additive sex, BO, reproduction, and area effects) *p*(best structure from 2–9 above)
28. φ(s + BO*g*R) *p*(best)	φ(interactive BO, reproduction, and area effects with additive sex effect) *p*(best structure from 2–9 above)
29. φ(CP) *p*(best)	φ(change-point in 2004 using best of (.), (t) or (T) models) *p*(best structure from 2–9 above)
30. φ(spline) *p*(best)	φ(cubic spline with knot midway between start year and 2002 and second knot at 2002) *p*(best structure from 2–9 above)
Study area surrogates	
31. φ(OWN) *p*(best)	Replace area effect in lowest QAIC$_c$ model from 9–29 with ownership effect
32. φ(ECO) *p*(best)	Replace area effect in lowest QAIC$_c$ model from 9–29 with ecoregion effect
33. φ(OWN*ECO) *p*(best)	Replace area effect in lowest QAIC$_c$ model from 9–29 with ownership and ecological region effects with interactions
34. φ(LAT) *p*(best)	Replace area effect in lowest QAIC$_c$ model from 9–29 with latitude effect
Habitat	
35. φ(s + g + [WA = OR + CA] *HAB1) *p*(best)	Sex included only if important in 1–34. Additive effects of area and habitat in WA and OR with minimum QAIC$_c$ model replacing habitat for CA. *p*(best structure from 2–9 above
36. φ(s + g HAB1) *p*(best)	Sex included only if important in 1–34. Interaction between area and HAB1. *p*(best structure from 2–9 above)
Climate	
37. φ(s + g + SOI + PDO) *p*(best)	φ(additive sex, area, Southern Oscillation Index, and Pacific Decadal Oscillation. *p*(best structure from 2–9 above)
38. φ[s + (g*SOI) + (g*PDO)] *p*(best)	φ(interaction between area and Southern Oscillation Index and area and Pacific Decadal Oscillation, with additive sex effects) *p*(best structure from 2–9 above)
39. φ(s + g + ENP) *p*(best)	φ(additive sex, area, and precipitation during early nesting season) *p*(best structure from 2–9 above)
40. φ(s + g*ENP) *p*(best)	φ(interaction between area and precipitation during early nesting season with additive sex effect) *p*(best structure from 2–9 above)
41. φ(s + g + ENT) *p*(best)	φ(additive sex, area, and temperature during early nesting season) *p*(best structure from 2–9 above)
42. φ(s + g*ENT) *p*(best)	φ(interaction between area and temperature during early nesting season with additive sex effect) *p*(best structure from 2–9 above)

APPENDIX G (*continued*)

Model	Description of Model Structure
Habitat-climate interactions	
43. φ(best habitat + best climate) p(best structure from 2–9 above)	φ(combine best habitat model from 35–36 with best climate model form 37–42 in additive model) p(best structure from 2–9 above)
44. φ(best habitat*best climate) p(best structure from 2–9 above)	φ(combine best habitat model from 35–36 with best climate model form 37–42 in interactive model) p(best structure from 2–9 above)

[a] When reproduction (R) appears as a covariate on recapture, it refers to the effect of reproduction in year t on recapture in year t. When R appears as a covariate on survival, it refers to the effect of reproduction in year t on survival in year $t + 1$.

APPENDIX H
Models used in the meta-analysis of λ of Northern Spotted Owls in Washington, Oregon, and California.

Model form was the apparent survival and recruitment parameterization. Model notation for random effects (RE) models includes the general model on which the random effects model is based. The last six models at the bottom of the list were developed *a posteriori* after looking at the ranking of the *a priori* models.

Model structure[a]
$\varphi(g^{*}t)\ p(g^{*}t)\ f(g^{*}t)$: RE φ(ECO) f(ECO)
$\varphi(g^{*}t)\ p(g^{*}t)\ f(g^{*}t)$: RE φ(g + BO) f(BO)
$\varphi(g^{*}t)\ p(g^{*}t)\ f(g^{*}t)$: RE φ(ECO) f(OWN + ECO)
$\varphi(g^{*}t)\ p(g^{*}t)\ f(g^{*}t)$: RE φ(g + BO) f(g + BO)
$\varphi(g^{*}t)\ p(g^{*}t)\ f(g^{*}t)$: RE φ(g + BO) f(g*BO)
$\varphi(g^{*}t)\ p(g^{*}t)\ f(g^{*}t)$: RE φ(g) f(g)
$\varphi(g^{*}t)\ p(g^{*}t)\ f(g^{*}t)$: RE φ(g) f(g + TT)
$\varphi(g^{*}t)\ p(g^{*}t)\ f(g^{*}t)$: RE φ(g + PDSI) f(g + ENP + ENT)
$\varphi(g^{*}t)\ p(g^{*}t)\ f(g^{*}t)$: RE φ(g) f(g + T)
$\varphi(g^{*}t)\ p(g^{*}t)\ f(g^{*}t)$: RE φ(g + PDSI) f(g + LNP)
$\varphi(g^{*}t)\ p(g^{*}t)\ f(g^{*}t)$: RE φ(g + PDSI) f(g + PDSI)
$\varphi(g^{*}t)\ p(g^{*}t)\ f(g^{*}t)$: RE φ(g + PDSI) f(g + SOI + PDO)
$\varphi(g^{*}t)\ p(g^{*}t)\ f(g^{*}t)$: RE φ(g) f(g*T)
$\varphi(g^{*}t)\ p(g^{*}t)\ f(g^{*}t)$: RE φ(g*T) f(g)
$\varphi(g^{*}t)\ p(g^{*}t)\ f(g + t)$
$\varphi(g^{*}t)\ p(g^{*}t)\ f(g^{*}t)$: RE φ(g + PDSI) f(g*LNP)
$\varphi(g^{*}t)\ p(g^{*}t)\ f(g^{*}t)$: RE φ(g + PDSI) f(g*PDSI)
$\varphi(g^{*}t)\ p(g^{*}t)\ f(g^{*}t)$: RE φ(g) f(g*TT)
$\varphi(g^{*}t)\ p(g^{*}t)\ f(g^{*}t)$: RE φ(g + PDSI) f(g*ENP + g*ENT)
$\varphi(g^{*}t)\ p(g^{*}t)\ f(g^{*}t)$: RE φ(g + PDSI) f(g*SOI + g*PDO)
$\varphi(g^{*}t)\ p(g^{*}t)\ f(g^{*}t)$: RE φ(g*HAB2) f(g+HAB2 + HAB3)
$\varphi(g^{*}t)\ p(g^{*}t)\ f(g^{*}t)$: RE φ(g*HAB2) f(g*HAB3)
$\varphi(g^{*}t)\ p(g^{*}t)\ f(g^{*}t)$: RE φ(g)
$\varphi(g^{*}t)\ p(g^{*}t)\ f(g^{*}t)$: RE φ(g*HAB2) f(g*HAB2 + g*HAB3)
$\varphi(g^{*}t)\ p(g^{*}t)\ f(g^{*}t)$: RE φ(g*TT)
$\varphi(g^{*}t)\ p(g^{*}t)\ f(g^{*}t)$: RE φ(ECO)
$\varphi(g^{*}t)\ p(g^{*}t)\ f(g^{*}t)$: RE φ(g + BO)
$\varphi(g^{*}t)\ p(g^{*}t)\ f(g^{*}t)$: RE φ(g*HAB2) f(g + HAB2)
$\varphi(g^{*}t)\ p(g^{*}t)\ f(g^{*}t)$: RE φ(g + PDSI)
$\varphi(g^{*}t)\ p(g^{*}t)\ f(g^{*}t)$: RE φ(BO)
$\varphi(g^{*}t)\ p(g^{*}t)\ f(g^{*}t)$: RE φ(OWN + ECO)
$\varphi(g^{*}t)\ p(g^{*}t)\ f(g^{*}t)$: RE φ(LAT)
$\varphi(g^{*}t)\ p(g^{*}t)\ f(g^{*}t)$: RE φ(g + T)
$\varphi(g^{*}t)\ p(g^{*}t)\ f(g^{*}t)$: RE φ(OWN)
$\varphi(g^{*}t)\ p(g^{*}t)\ f(g^{*}t)$: RE φ(g*PDSI)

APPENDIX H *(continued)*

Model structure[a]

$\varphi(g^*t)\ p(g^*t)\ f(g^*t)$: RE $\varphi(g+SOI + PDO)$

$\varphi(g^*t)\ p(g^*t)\ f(g^*t)$: RE $\varphi(g^*T)$

$\varphi(g^*t)\ p(g^*t)\ f(g^*t)$: RE $\varphi(g^*BO)$

$\varphi(g^*t)\ p(g^*t)\ f(g^*t)$: RE $\varphi(g + ENP + ENT)$

$\varphi(g^*t)\ p(g^*t)\ f(g^*t)$: RE $\varphi(g^*SOI + g^*PDO)$

$\varphi(g^*t)\ p(g^*t)\ f(g^*t)$: RE $\varphi(g^*HAB2)$

$\varphi(g^*t)\ p(g^*t)\ f(g^*t)$: RE $\varphi(g^*ENP + g^*ENT)$

$\varphi(g^*t)\ p(g^*t)\ f(g^*t)$: RE $\varphi(g + HAB2)$

$\varphi(g^*t)\ p(g^*t)\ f(g^*t)$: RE $\varphi(g + TT)$

$\varphi(g^*t)\ p(g^*t)\ f(g^*t)$

$\varphi(g^*t)\ p(g^*t)\ f(g^*t)$: RE $\varphi(ECO + BO)\ f(ECO)$

$\varphi(g^*t)\ p(g^*t)\ f(g^*t)$: RE $\varphi(ECO + BO)\ f(ECO + BO)$

$\varphi(g^*t)\ p(g^*t)\ f(g^*t)$: RE $\varphi(ECO)\ f(ECO^*BO)$

$\varphi(g^*t)\ p(g^*t)\ f(g^*t)$: RE $\varphi(ECO^*BO)\ f(ECO^*BO)$

$\varphi(g^*t)\ p(g^*t)\ f(g^*t)$: RE $\varphi(ECO)\ f(ECO + BO)$

$\varphi(g^*t)\ p(g^*t)\ f(g^*t)$: RE $\varphi(ECO + BO)\ f(ECO)$

[a] Model notation indicates structure for effects of study area (g), time (t), linear time trend (T), quadratic time trend (TT), ecoregion (ECO), proportion of territories with Barred Owl detections (BO), land ownership (OWN), early nesting season precipitation (ENP), early nesting season temperature (ENT), Palmer Drought Severity Index (PDSI), late nesting season precipitation (LNP), Southern Oscillation Index (SOI), Pacific Decadal Oscillation (PDO), percent cover of suitable owl habitat within 2.4 km of owl activity centers used in λ analysis (HAB2), percent cover of suitable owl habitat within 23 km of owl activity centers used in λ analysis, minus the area of HAB2 (HAB3).

LITERATURE CITED

Akaike, H. 1973. Information theory as an extension of the maximum likelihood principle. Pp. 267–281 *in* B. N. Petrov and F. Csaki (editors). Second international symposium on information theory. Akademiai Kiado, Budapest, Hungary.

Akaike, H. 1985. Prediction and entropy. Pp. 1–24 *in* A. C. Atkinson and S. E. Fienberg (editors). A celebration of statistics: The ISI centenary volume. Springer-Verlag, New York, NY.

Alley, W. M. 1984. The Palmer Drought Severity Index: Limitations and assumptions. Journal of Climate and Applied Meteorology 23:1100–1109.

Anderson, D. R., and K. P. Burnham. 1992. Demographic analysis of Northern Spotted Owl populations. Pp. 66–75 *in* USDI Fish and Wildlife Service. Final draft recovery plan for the Northern Spotted Owl. Volume 2. USDI Fish and Wildlife Service, Region 1, Portland, OR.

Anderson, D. R., K. P. Burnham, A. B. Franklin, R. J. Gutiérrez, E. D. Forsman, R. G. Anthony, G. C. White, and T. M. Shenk. 1999. A protocol for conflict resolution in analyzing empirical data related to natural resource controversies. Wildlife Society Bulletin 27:1050–1058.

Anderson, D. R., K. P. Burnham, and G. C. White. 1994. AIC model selection in overdispersed capture-recapture data. Ecology 75:1780–1793.

Anthony, R. G., E. D. Forsman, A. B. Franklin, D. R. Anderson, K. P. Burnham, G. C. White, C. J. Schwarz, J. D. Nichols, J. E. Hines, G. S. Olson, S. H. Ackers, L. W. Andrews, B. L. Biswell, P. C. Carlson, L. V. Diller, K. M. Dugger, K. E. Fehring, T. L. Fleming, R. P Gearhardt, S. A. Gremel, R. J. Gutiérrez, P. J. Happe, D. R. Herter, J. M. Higley, R. B. Horn, L. L. Irwin, P. J. Loschl, J. A. Reid, and S. G. Sovern. 2006. Status and trends in demography of Northern Spotted Owls, 1985–2003. Wildlife Monographs No. 163.

Astheimer, L. B., W. A. Buttemer, and J. C. Wingfield. 1995. Seasonal and acute changes in adrenocortical responsiveness in an arctic-breeding bird. Hormonal Behavior 29:442–457.

Bailey, L. L., J. A. Reid, E. D. Forsman, and J. D. Nichols. 2009. Modeling co-occurrence of Northern Spotted and Barred Owls: Accounting for detection probability differences. Biological Conservation 142:2983–2989.

Bart, J., and E. D. Forsman 1992. Dependence of Northern Spotted Owls *Strix occidentalis caurina* on old-growth forests in the western USA. Biological Conservation 62:95–100.

Blakesley, J. A., B. R. Noon, and D. W. H. Shaw. 2001. Demography of the California Spotted Owl in northeastern California. Condor 103:667–677.

Boyce, M. S., L. L. Irwin, and R. Barker. 2005. Demographic meta-analysis: Synthesizing vital rates for Spotted Owls. Journal of Applied Ecology 42:38–49.

Bradley, M., R. Johnstone, G. Court, and T. Duncan. 1997. Influence of weather on breeding success of Peregrine Falcons in the Arctic. Auk 114:786–791.

Brownie, C., D. R. Anderson, K. P. Burnham, and D. S. Robson. 1978. Statistical inference from band recovery data—a handbook. USDI Fish and Wildlife Service Resource Publication 131.

Buchanan, J. B., R. J. Gutiérrez, R. G. Anthony, T. Cullinan, L. V. Diller, E. D. Forsman, and A. B. Franklin. 2007. A synopsis of suggested approaches to address potential competitive interactions between Barred Owls (*Strix varia*) and Spotted Owls (*S. occidentalis*). Biological Invasions 9:679–691.

Buchanan, J.G., L.L. Irwin, and E.L. McCutchen. 1995. Within-stand nest site selection by Spotted

Owls in the eastern Washington Cascades. Journal of Wildlife Management 59:301–310.

Burnham, K. P., and D. R. Anderson. 2002. Model selection and multimodel inference: A practical information-theoretical approach. 2nd ed. Springer-Verlag, New York, NY.

Burnham, K. P., D. R. Anderson, and G. C. White. 1994. Estimation of vital rates of the Northern Spotted Owl. Pp. 1–26 in Appendix J of the Final supplemental environmental impact statement on management of habitat for late-successional and old-growth forest related species within the range of the Northern Spotted Owl. Volume 2. USDA Forest Service and USDI Bureau of Land Management, Portland, OR.

Burnham, K. P., D. R. Anderson, and G. C. White. 1996. Meta-analysis of vital rates of the Northern Spotted Owl. Pp. 92–101 in E. D. Forsman, S. DeStefano, M. G. Raphael, and R. J. Gutiérrez (editors), Demography of the Northern Spotted Owl. Studies in Avian Biology No. 17.

Burnham, K. P., D. R. Anderson, G. C. White, C. Brownie, and K. H. Pollock. 1987. Design and analysis of fish survival experiments based on release-recapture. American Fisheries Society Monograph 5:1–412.

Burnham, K. P., and G. C. White. 2002. Evaluation of some random effects methodology applicable to bird ringing data. Journal of Applied Statistics 29:245–264.

Carey, A. B., S. P. Horton, and B. L. Biswell. 1992. Northern Spotted Owls: Influence of prey base and landscape character. Ecological Monographs 62:223–250.

Carrete, M., J. A. Sanchez-Zapata, J. F. Calvo, and R. Lande. 2005. Demography and habitat availability in territorial occupancy of two competing species. Oikos 108:125–136.

Caswell, H. 2001. Matrix population models: Construction, analysis and interpretation. 2nd ed. Sinauer Associates, Sunderland, MA.

Cichoń, M., P. Olejniczak, and L. Gustaffson. 1998. The effect of body condition on the cost of reproduction in female Collared Flycatchers Ficedula albicollis. Ibis 140:128–10.

Clobert, J., C. M. Perrins, R. H. McCleery, and A. G. Gosler. 1988. Survival rate in the Great Tit Parus major in relation to sex, age, and immigration status. Journal of Animal Ecology 57:287–306.

Clutton-Brock, T. H., (editor). 1988. Reproductive success. Studies of individual variation in contrasting breeding systems. University of Chicago Press, Chicago, IL.

Cohen, W. B., M. Fiorella, J. Gray, E. Helmer, and K. Anderson. 1998. An efficient and accurate method for mapping forest clearcuts in the Pacific Northwest using Landsat imagery. Photogrammetric Engineering and Remote Sensing 64:293–300.

Cormack, R. M. 1964. Estimates of survival from the sighting of marked animals. Biometrika 51:429–438.

Courtney, S. P., J. A. Blakesley, R. E. Bigley, M. L. Cody, J. P. Dumbacher, R. C. Fleischer, A. B. Franklin, J. F. Franklin, R. J. Gutiérrez, J. M. Marzluff, and L. Sztukowski. 2004. Scientific evaluation of the status of the Northern Spotted Owl. Sustainable Ecosystems Institute, Portland, OR.

Croxall, J. P., P. Rothery, S. P. C. Pickering, and P. A. Prince. 1990. Reproductive performance, recruitment and survival of Wandering Albatrosses Diomedea exulans at Bird Island, South Georgia. Journal of Animal Ecology 59:775–796.

Daly, C. 2006. Guidelines for assessing the suitability of spatial climate data sets. International Journal of Climatology 26:707–721.

Dark, S. J., R. J. Gutiérrez, and G. I. Gould, Jr. 1998. The Barred Owl (Strix varia) invasion in California. Auk 115:50–56.

Davis, R., and J. B. Lint. 2005. Habitat status and trend. Pp. 21–82 in J. B. Lint (technical coordinator). Northwest Forest Plan—the first ten years (1994–2003): Status and trend of Northern Spotted Owl populations and habitat. USDA Forest Service General Technical Report PNW-GRT-648.

Davis, R. J., and J. B. Lint. 2006. Appendix F. Comparison of habitat conditions for Spotted Owls in 14 demographic study areas to conditions on federal lands surrounding them. Pp. 44–47 in Anthony et al. Status and trends in demography of Northern Spotted Owls, 1985–2003. Wildlife Monographs No. 163:1–48.

Dietrich, W. 2003. The final forest. The Mountaineers Press, Seattle, WA.

Diller, L. V., and D. M. Thome. 1999. Population density of Northern Spotted Owls in managed young-growth forests in coastal northern California. Journal of Raptor Research 33:275–286.

Dugger, K. M., F. Wagner, R. G. Anthony, and G. S. Olson. 2005. The relationship between habitat characteristics and demographic performance of Northern Spotted Owls in southern Oregon. Condor 107:863–878.

Dugger, K. M., R. G. Anthony, and L. S. Andrews. In press. Transient dynamics of invasive competition: Barred Owls, Spotted Owls, and the demons of competition present. Ecological Applications.

Durbin, K. 1996. Tree huggers: Victory, defeat and renewal in the northwest ancient forest campaign. The Mountaineers Press, Seattle, WA.

Ervin, K. 1989. Fragile majesty. The Mountaineers, Seattle Press, WA.

Evans, M., N. Hastings, and B. Peacock. 1993. Statistical distributions. 2nd ed. J. Wiley and Sons, New York, NY.

Fleming, T. L., J. L. Halverson, and J. B. Buchanan. 1996. Use of DNA analysis to identify sex of

Northern Spotted Owls (*Strix occidentalis caurina*). Journal of Raptor Research 30:118–122.

Forsman, E. D. 1981. Molt of the Spotted Owl. Auk 98:735–742.

Forsman, E. D. 1983. Methods and materials for locating and studying Spotted Owls. USDA Forest Service General Technical Report PNW-162. USDA Forest Service, Pacific Northwest Research Station, Portland, OR.

Forsman, E. D., R. G. Anthony, J. A. Reid, P. J. Loschl, S. G. Sovern, M. Taylor, B. L. Biswell, A. Ellingson, E. C. Meslow, G. S. Miller, K. A. Swindle, J. A. Thrailkill, F. F. Wagner, and D. E. Seaman. 2002. Natal and breeding dispersal of Northern Spotted Owls. Wildlife Monographs 149:1–35.

Forsman, E. D., S. DeStefano, M. G. Raphael, and R. J. Gutiérrez (editors). 1996a. Demography of the Northern Spotted Owl. Studies in Avian Biology 17:1–122.

Forsman E. D., A. B. Franklin, F. M. Oliver, and J. P. Ward. 1996b. A color band for Spotted Owls. Journal of Field Ornithology 67:507–510.

Forsman, E. D., T. J. Kaminski, J. C. Lewis, K. J. Maurice, and S. G. Sovern. 2005. Home range and habitat use of Northern Spotted Owls on the Olympic Peninsula, Washington. Journal of Raptor Research 39:365–377.

Forsman, E. D., E .C. Meslow, and H. M. Wight. 1984. Distribution and biology of the Spotted Owl in Oregon. Wildlife Monographs 87:1–64.

Franklin, A. B. 1992. Population regulation in Northern Spotted Owls: Theoretical implications for management. Pp. 815–830 *in* D. R. McCullough and R. H. Barrett (editors). Wildlife 2001: Populations. Elsevier Applied Science, London, UK.

Franklin, A. B. 2001. Exploring ecological relationships in survival and estimating rates of population change using program MARK. Pp. 350–356 *in* R. Field, R. J. Warren, H. K. Okarma, and P. R. Sievert (editors). Wildlife, Land, and People: Priorities for the 21st century. The Wildlife Society, Bethesda, MD.

Franklin, A. B., D. R. Anderson, E. D. Forsman, K. P. Burnham, and F. W. Wagner. 1996. Methods for collecting and analyzing demographic data on the Northern Spotted Owl. Pp. 12–20 *in* Forsman, E. D., S. DeStefano, M. G. Raphael, and R. J. Gutiérrez (editors). Demography of the Northern Spotted Owl. Studies in Avian Biology No. 17.

Franklin, A. B., D. R. Anderson, R. J. Gutiérrez, and K. P. Burnham. 2000. Climate, habitat quality, and fitness in Northern Spotted Owl populations in northwestern California. Ecological Monographs 70:539–590.

Franklin, A. B., K. P. Burnham, G. C. White, R. G. Anthony, E. D. Forsman, C. Schwarz, J. D. Nichols, and J. E. Hines. 1999. Range-wide status and trends in Northern Spotted Owl populations. USGS Colorado Cooperative Fish and Wildlife Research Unit, Colorado State University, Fort Collins, CO.

Franklin, A. B., R. J. Gutiérrez, J. D. Nichols, M. E. Seamans, G. C. White, G. S. Zimmerman, J. E. Hines, T. E. Munton, W. S. LaHaye, J. A. Blakesley, G. N. Steger, B. R. Noon, D. W. H. Shaw, J. J. Keane, T. L. McDonald, and S. Britting. 2004. Population dynamics of the California Spotted Owl (*Strix occidentalis occidentalis*): A meta-analysis. Ornithological Monographs 54:1–54.

Franklin, A. B., J. D. Nichols, R. G. Anthony, K. P. Burnham, G. C. White, E. D. Forsman, and D. R. Anderson. 2006. Comment on "Are survival rates of Northern Spotted Owls biased." Canadian Journal of Zoology 84:1375–1379.

Franklin, J. F., and C. T. Dyrness. 1973. Natural vegetation of Oregon and Washington. USDA Forest Service General Technical Report PNW-8. USDA Forest Service, Pacific Northwest Forest and Range Experiment Station, Portland, OR.

Franklin, J. F., and T. A. Spies. 1991. Ecological definitions of old-growth Douglas-fir forests. Pp. 61–69 *in* Ruggiero, L. F., K. B. Aubry, A. B. Carey, and M. H. Huff (technical coordinators). Wildlife and vegetation of unmanaged Douglas-fir forests. USDA Forest Service General Technical Report PNW-GTR-285. USDA Forest Service, Pacific Northwest Research Station, Portland, OR.

Gaillard, J.-M., M. Festa-Bianchet, and N. G. Yoccoz. 1998. Population dynamics of large herbivores: Variable recruitment with constant adult survival. Trends in Ecology and Evolution 13:58–63.

Gaillard, J.-M., M. Festa-Bianchet, N. G. Yoccoz, A. Loison, and C. Toigo. 2000. Temporal variation in fitness components and population dynamics of large herbivores. Annual Review of Ecology and Systematics. 31:367–393.

Gaillard, J.-M., and N. G. Yoccoz. 2003. Temporal variation in the survival of mammals: A case of environmental canalization? Ecology 84:3294–3306.

Glenn, E. M. 2009. Local weather, regional climate, and population dynamics of Northern Spotted Owls in Washington and Oregon. PhD Dissertation, Oregon State University, Corvallis, OR.

Glenn, E. M., R. G. Anthony, and E. D. Forsman. 2010. Population trends in Northern Spotted Owls: Associations with climate in the Pacific Northwest. Biological Conservation 143:2543–2552.

Glenn, E. M., R. G. Anthony, E. D. Forsman, and G. S. Olson. 2011. Local weather, regional climate, and annual survival of the Northern Spotted Owl. Condor 113:159–176.

Glenn, E. M., R. G. Anthony, E. D. Forsman, and G. S. Olson. In press. Reproduction of Northern Spotted Owls: The role of local weather and regional climate. Journal of Wildlife Management.

Glenn, E. M., M. C. Hansen, and R. G. Anthony. 2004. Spotted Owl home-range and habitat use in young forests of western Oregon. Journal of Wildlife Management 68:33–50.

Green, P. J., and B. W. Silverman. 1994. Nonparametric regression and generalized linear models. A roughness penalty approach. Chapman & Hall, London, UK.

Hakkarainen, H., V. Koivunen, and E. Korpimäki. 1997. Reproductive success and parental effort of Tengmalm's Owls: Effects of spatial and temporal variation in habitat quality. Ecoscience 4:35–42.

Hamer, T. E., E. D. Forsman, and E. M. Glenn. 2007. Home range attributes and habitat selection of Barred Owls and Spotted Owls in an area of sympatry. Condor 109:750–768.

Hastie, J. J., and R. J. Tibshirani. 1990. Generalized additive models. Chapman & Hall, London, UK.

Healey, S. P., W. B. Cohen, T. A. Spies, M. Moeur, D. Pflugmacher, M. G. Whitley, and M. Lefsky. 2008. The relative impact of harvest and fire upon landscape-level dynamics of older forests: Lessons from the Northwest Forest Plan. Ecosystems 11:1106–1119

Hershey, K. T., E. C. Meslow, and F. L. Ramsey. 1998. Characteristics of forests at Spotted Owl nest sites in the Pacific Northwest. Journal of Wildlife Management 62:1398–1410.

Herter, D. R., L. L. Hicks, H. C. Stabins, J. J. Millspaugh, and A. J. Stabins. 2002. Roost site characteristics of Northern Spotted Owls in the nonbreeding season in central Washington. Forest Science 48:437–444.

Hines, J. E., and J. D. Nichols. 2002. Investigations of potential bias in the estimation of λ using Pradel's (1996) model for capture-recapture data. Journal of Applied Statistics 29:573–587.

Horn, H. S., and R. H. MacArthur. 1972. Competition among fugitive species in a harlequin environment. Ecology 53:749–752.

Hurvich, C. M., and C.-L. Tsai. 1989. Regression and time series model selection in small samples. Biometrika 76:297–307.

Hwang, W.-D., and A. Chao. 1995. Quantifying the effects of unequal catchabilities on Jolly-Seber estimators via sample coverage. Biometrics 51:128–141.

Johnson, D. H., G. C. White, A. B. Franklin, L. V. Diller, I. Blackburn, D. J. Pierce, G. S. Olson, J. B. Buchanan, J. Thrailkill, B. Woodbridge, and M. Ostwald. 2008. Study designs for Barred Owl removal experiments to evaluate effects on Northern Spotted Owls. USDI Fish and Wildlife Service, Portland, OR.

Jolly, G. M. 1965. Explicit estimates from capture-recapture data with both death and immigration—stochastic model. Biometrika 52:225–247.

Karell, P., K. Ahola, T. Karstinen, A. Zolei, and J. E. Brommer. 2009. Population dynamics in a cyclic environment: consequences of cyclic food abundance on Tawny Owl reproduction and survival. Journal of Animal Ecology 78:1050–1062.

Kelly, E. G. 2001. The range expansion of the Northern Barred Owl: an evaluation of the impact on Spotted Owls. M.S. Thesis, Oregon State University, Corvallis, OR.

Kelly, E. G., E. D. Forsman, and R. G. Anthony. 2003. Are Barred Owls displacing Spotted Owls? Condor 105:45–53.

Korpimäki, E. 1992. Fluctuating food abundance determines the lifetime reproductive success of male Tengmalm's Owls. Journal of Animal Ecology 61:103–111.

Küchler, A. W. 1977. The map of the natural vegetation of California. Pp. 909–938 in M. G. Barbour and J. Major (editors). Terrestrial vegetation of California. J. Wiley and Sons, New York, NY.

LaHaye, W. S., and R. J. Gutiérrez. 1999. Nest sites and nesting habitat of the Northern Spotted Owl in northwestern California. Condor 101:324–340.

LaHaye, W. S., R. J. Gutiérrez, and D. R. Call. 1992. Demography of an insular population of Spotted Owls (Strix occidentalis occidentalis). Pp. 803–814 in D. R. McCullough and R. H. Barrett (editors). Wildlife 2001: populations. Elsevier Applied Science, London, UK.

LaHaye, W. S., R. J. Gutiérrez, and D. R. Call. 1997. Nest-site selection and reproductive success of California Spotted Owls. Wilson Bulletin 109:42–51.

LaHaye, W. S., G. S. Zimmerman, and R. J. Gutiérrez. 2004. Temporal variation in the vital rates of an insular population of Spotted Owls (Strix occidentalis occidentalis): Contrasting effects of weather. Auk 121:1056–1069.

Lande, R. 1988. Demographic models of the Northern Spotted Owl (Strix occidentalis caurina). Oecologia 75:601–607.

Lande, R. 1991. Population dynamics and extinction in heterogeneous environments: The Northern Spotted Owl. Pp. 566–580 in C. M. Perrins, J.-D. Lebreton, and G. J. M. Hirons (editors). Bird population studies: Relevance to conservation and management. Oxford University Press, Oxford, UK.

Lebreton, J.-D., K. P. Burnham, J. Clobert, and D. R. Anderson. 1992. Modeling survival and testing biological hypotheses using marked animals: a unified approach with case studies. Ecological Monographs 62:67–118.

Lehmkuhl, J. F., K. D. Kistler, and J. S. Begley. 2006a. Bushy-tailed woodrat abundance in dry forests of eastern Oregon. Journal of Mammalogy 87:371–379.

Lehmkuhl, J. F., K. D. Kistler, J. S. Begley, and J. Boulanger. 2006b. Demography of northern flying squirrels informs ecosystem management of western interior forests. Ecological Applications 16:584–600.

Levins, R., and D. Culver. 1971. Regional coexistence of species and competition between rare species.

Proceeding of the National Academy of Science 68:1246–1248.

Lint, J., B. Noon, R. Anthony, E. Forsman, M. Raphael, M. Collopy, and E. Starkey. 1999. Northern Spotted Owl effectiveness monitoring plan for the northwest forest plan. USDA Forest Service General Technical Report PNW-GTR-440.

Lint, J. (technical coordinator). 2005. Northwest forest plan—the first 10 years (1994–2003): Status and trends of Northern Spotted Owl populations and habitat. USDA Forest Service General Technical Report PNW-GTR-648.

Loehle, C., L. Irwin, D. Rock, and S. Rock. 2005. Are survival rates for Northern Spotted Owls biased? Canadian Journal of Zoology 83:1386–1390.

Manly, B. F. J., L. L. McDonald, and T. L. McDonald. 1999. The robustness of mark-recapture methods: A case study for the Northern Spotted Owl. Journal of Agricultural, Biological, and Environmental Statistics 4:78–101.

Martin, K. 1995. Patterns and mechanisms for age-dependent reproduction and survival in birds. American Zoologist 35:340-348.

McCleery, R. H., J. Clobert, R. Julliard, and C. M. Perrins. 1996. Nest predation and delayed cost of reproduction in the Great Tit. Journal of Animal Ecology 65:96–104.

McDonald, P. G., P. D. Olsen, and A. Cockburn. 2004. Weather dictates reproductive success and survival in the Australian Brown Falcon *Falco berigora*. Journal of Animal Ecology 73:683–692.

Millon, A., S. J. Petty, and X. Lambin. 2009. Pulsed resources affect the timing of first breeding and lifetime reproductive success of Tawny Owls. Journal of Animal Ecology 79:426–435.

Moen, C. A., A. B. Franklin, and R. J. Gutiérrez. 1991. Age determination of subadult Northern Spotted Owls in northwest California. Wildlife Society Bulletin 19:489–493.

Møller, A. P., and T. Szép. 2002. Survival rate of adult Barn Swallows *Hirundo rustica* in relation to sexual selection and reproduction. Ecology 83:2220–2228.

Newton, I., (editor). 1989. Lifetime reproduction in birds. Academic Press, San Diego, CA.

NOAA. 2008a. NOAA Satellite and Information Service and the National Climatic Data Center. <http://www.ncdc.noaa.gov/oa/climate/onlineprod/drought/xmgr.html#gr> (20 October 2008).

NOAA. 2008b. NOAA Satellite and Information Service and the National Climatic Data Center. <http://www.cpc.ncep.noaa.gov/data/indices/soi> (15 October 2008).

Noon, B. R., and C. M. Biles. 1990. Mathematical demography of Spotted Owls in the Pacific Northwest. Journal of Wildlife Management 54:18–27.

Olsen, P. D., and J. Olsen. 1989. Breeding of the Peregrine Falcon *Falco peregrinus*. III. Weather, nest quality and the timing of egg laying. Emu 89:6–14.

Olson, G. S., E. M. Glenn, R. G. Anthony, E. D. Forsman, J. A. Reid, P. J. Loschl, and W. J. Ripple. 2004. Modeling demographic performance of Northern Spotted Owls relative to forest habitat in Oregon. Journal of Wildlife Management 68:1039–1053.

Olson, G. S., R. G. Anthony, E. D. Forsman, S. H. Ackers, P. J. Loschl, J. A. Reid, K. M. Dugger, E. M. Glenn, and W. J. Ripple. 2005. Modeling of site occupancy dynamics for Northern Spotted Owls, with emphasis on the effects of Barred Owls. Journal of Wildlife Management 69:918–932.

Oregon Climate Service. 2008. Parameter-elevation regressions on independent slopes models (PRISM) <http://www.ocs.oregonstate.edu/prism> (20 October 2008).

Parson, E. A., P. W. Mote, A. Hamlet, N. Mantua, A. Snover, W. Keeton, E. Miles, D. Canning, and K. G. Ideker. 2001. Potential consequences of climate variability and change for the Pacific Northwest. Chapter 9 *in* Climate change impacts on the United States: The potential consequences of climate variability and change, report of the National Assessment Synthesis Team, US Global Change Research Program, Washington D.C.

Pfister, C. A. 1998. Patterns of variation in stage-structured populations: evolutionary predictions and ecological implications. Proceedings of the National Academy of Science 95:213–218.

Pollock, K. H., J. D. Nichols, C. Brownie, and J. E. Hines. 1990. Statistical inference for capture-recapture experiments. Wildlife Monograph 107:1–97.

Pradel, R. 1996. Utilization of capture-mark-recapture for the study of recruitment and population growth rate. Biometrics 52:703–709.

Pyle, P., N. Nur, W. Sydeman, and S. Emslie. 1997. Cost of reproduction and the evolution of deferred breeding in the Western Gull. Behavioral Ecology 8:140–147.

Ramsey, F. L, and D. W. Schafer. 2002. The statistical sleuth: A course in methods of data analysis. 2nd ed. Duxbury, Thompson Learning, Inc., Pacific Grove, CA.

Raphael, M. G., R. G. Anthony, S. DeStefano, E. D. Forsman, A. B. Franklin, R. Holthausen, E. C. Meslow, and B. R. Noon. 1996. Use, interpretation, and implications of demographic analyses of Northern Spotted Owl populations. Studies in Avian Biology 17:102–112.

Reid, J. A., R. B. Horn, and E. D. Forsman. 1999. Detection rates of Spotted Owls based on acoustic-lure and live-lure surveys. Wildlife Society Bulletin 27:986–990.

Roberge, J., and P. Angelstam. 2004. Usefulness of the umbrella species concept as a conservation tool. Conservation Biology 18:76–85.

Rogers, C. M., M. Ramenofsky, E. D. Ketterson, V. Nolan, Jr., and J. C. Wingfield. 1993. Plasma corticosterone, adrenal mass, winter weather, and season in nonbreeding populations of Dark-eyed Juncos (*Junco hyemalis hyemalis*). Auk 110:279–285.

Rohner, C. 1996. The numerical response of Great Horned Owls to the snowshoe hare cycle: Consequences of non-territorial "floaters" on demography. Journal of Animal Ecology 65:359–370.

Romero, L. M., J. M. Reed, and J. C. Wingfield. 2000. Effects of weather on corticosterone responses in wild free-living passerine birds. General and Comparative Endocrinology 118:113–122.

Rosenberg, D. K., and R. G. Anthony. 1992. Characteristics of northern flying squirrel populations in young second- and old-growth forests in western Oregon. Canadian Journal of Zoology 70:161–166.

Rosenberg, D. K., K. A. Swindle, and R. G. Anthony. 2003. Influence of prey abundance on Northern Spotted Owl reproductive success in western Oregon. Canadian Journal of Zoology 81:1715–1725.

Rotella, J. J., R. G. Clark, and A. D. Afton. 2003. Survival of female Lesser Scaup: Effects of body size, age, and reproductive effort. Condor 105:336–347.

Rotenberry J. T., and J. A. Wiens. 1991. Weather and reproductive variation in shrubsteppe sparrows: A hierarchical analysis. Ecology 72:1325–1335.

Saether, B. E. 1990. Age-specific variation in the reproductive performance of birds. Pp. 251–283 *in* D. M. Power (editor), Current Ornithology. Plenum Press, New York, NY.

Sakamoto, Y., M. Ishiguro, and G. Kitagawa. 1986. Akaike information criterion statistics. KTK Scientific Publishers, Tokyo, Japan.

Salafsky, S.R., R.T. Reynolds, and B.R. Noon. 2005. Patterns of temporal variation in Goshawk reproduction and prey resources. Journal of Raptor Research 29:237–246.

Sandercock, B. K., and S. R. Beissinger. 2002. Estimating rates of population change for a neotropical parrot with ratio, mark-recapture, and matrix methods. Journal of Applied Statistics 29:589–607.

SAS Institute, Inc. 2008. SAS/STAT® 9.2 users guide. SAS Institute, Inc., Cary, NC.

Saurola, P. 1987. Mate and nest site fidelity in Ural and Tawny Owls. Pp. 81–86 *in* R. W. Nero, R. J. Clarke, R. J. Knapton, and R. H. Hamre (editors). Biology and conservation of northern forest owls. USDA Forest Service, General Technical Report RM-142.

Saurola, P. 2003. Life of the Ural Owl *Strix uralensis* in a cyclic environment: Some results of a 36-year study. Avocetta 27:76–79.

Seamans, M. E., and R. J. Gutiérrez. 2007. Habitat selection in a changing environment: The relationship between habitat alteration and Spotted Owl occupancy and breeding dispersal. Condor 109:566–576.

Seamans, M. E., R. J. Gutiérrez, and C. A. May. 2002. Mexican Spotted Owl (*Strix occidentalis*) population dynamics: Influence of climatic variation on survival and reproduction. Auk 119:321–334.

Seamans, M. E., R. J. Gutiérrez, C. A. May, and M. Z. Peery. 1999. Demography of two Mexican Spotted Owl populations. Conservation Biology 13:744–754.

Seamans, M. E., R. J. Gutiérrez, C. A. Moen, and M. Z. Peery. 2001. Spotted Owl demography in the central Sierra Nevada. Journal of Wildlife Management 65:425–431.

Seber, G. A. F. 1965. A note on the multiple-recapture census. Biometrika 52:249–259.

Singleton, P. H., J. F. Lehmkuhl, W. L. Gaines, and S. A. Graham. 2010. Barred Owl space use and habitat selection in the eastern Cascades, Washington. Journal of Wildlife Management 74:285–294.

Smith, G. T., J. C. Wingfield, and R. R. Veit. 1994. Adrenocortical response to stress in the Common Diving Petrel, *Pelecanoides urinatrix*. Physiological Zoology 67:526–537.

Stearns, S. C. 1976. Life-history tactics: a review of ideas. Quarterly Review of Biology 51:3–47.

Steenhof, K., M. N. Kochert, and T. L. McDonald. 1997. Interactive effects of prey and weather on Golden Eagle reproduction. Journal of Animal Ecology 66:350–362.

Swindle, K. A., W. J. Ripple, E. C. Meslow, and D. Schafer. 1999. Old-forest distribution around Spotted Owl nests in the central Cascade Mountains, Oregon. Journal of Wildlife Management 63:1212–1221.

Tavecchia, G., R. Pradel, V. Boy, A. R. Johnson, and F. Cézilly. 2001. Sex- and age-related variation in survival and cost of first reproduction in Greater Flamingos. Ecology 82:165–174.

Thomas, J. W., E. D. Forsman, J. B. Lint, E. C. Meslow, B. R. Noon, and J. Verner. 1990. A conservation strategy for the Northern Spotted Owl: Report of the Interagency Scientific Committee to address the conservation of the Northern Spotted Owl. USDA Forest Service, USDI Bureau of Land Management, Fish and Wildlife Service, and National Park Service, Portland, OR.

Thomas, J. W., M. G. Raphael, R. G. Anthony, E. D. Forsman, A. G. Gunderson, R. S. Holthausen, B. G. Marcot, G. H. Reeves, J. R. Sedell, and D. M. Solis. 1993. Viability assessments and management considerations for species associated with late-successional and old-growth forests of the Pacific Northwest: Report of the Scientific Analysis Team. USDA Forest Service, Portland, OR.

Thome, D. M., C. J. Zabel, and L. V. Diller. 2000. Spotted Owl turnover and reproduction in managed forests of north-coastal California. Journal of Field Ornithology 71:140–146.

Tickell, W. L. N. 1968. The biology of the great albatrosses, *Diomedea exulans* and *Diomedea epomophora*. Pp. 1–55 *in* O. L. Austin, Jr. (editor). Antarctic bird studies. American Geophysical Union, Washington, D.C.

University of Washington. 2008. Joint Institute for the Study of the Atmosphere and Ocean. <http://jisao.washington.edu/pdo/PDO.latest> (15 October 2008).

USDA Forest Service and USDI Bureau of Land Management. 1994. Final supplemental environmental impact statement on management of habitat for late-successional and old-growth forest related species within the range of the Northern Spotted Owl. USDA Forest Service and USDI Bureau of Land Management, Portland, OR.

USDI Fish and Wildlife Service. 1990. Endangered and threatened wildlife and plants: Determination of threatened status for the Northern Spotted Owl. Federal Register 55:26114–26194.

USDI Fish and Wildlife Service. 1992. USDI Fish and Wildlife Service. Final draft recovery plan for the Northern Spotted Owl. Volumes 1–2. USDI Fish and Wildlife Service, Region 1, Portland, OR.

USDI Fish and Wildlife Service. 2008. Final recovery plan for the Northern Spotted Owl (*Strix occidentalis caurina*). USDI Fish and Wildlife Service, Portland, OR.

Van Deusen, P. C., L. L. Irwin, and T. L. Fleming. 1998. Survival estimates for the Northern Spotted Owl. Canadian Journal of Forest Research 28:1681–1685.

Venables, W. N., and B. D. Ripley. 1999. Modern applied statistics with S-plus. 3rd ed. Springer, New York, NY.

Wagner, F. F., E. C. Meslow, G. M. Bennett, C. J. Larson, S. M. Small, and S. DeStefano. 1996. Demography of Northern Spotted Owls in the southern Cascades and Siskiyou Mountains, Oregon. Studies in Avian Biology 17:67–76.

Ward, J. P., Jr., and W. M. Block. 1995. Mexican Spotted Owl prey ecology. Pp. 1–48 *in* USDI Fish and Wildlife Service, Recovery plan for the Mexican Spotted Owl. Volume 2. USDI Fish and Wildlife Service, Albuquerque, NM.

Weimerskirch, H., J. Clobert, and P. Jouventin. 1987. Survival in five southern albatrosses and its relationship with their life history. Journal of Animal Ecology 56:1043–1055.

White, G. C., and R. E. Bennetts. 1996. Analysis of frequency count data using the negative binomial distribution. Ecology 77:2549–2557.

White, G. C., and K. P. Burnham. 1999. Program MARK: Survival estimation from populations of marked animals. Bird Study 46:120–138.

White, G. C., K. P. Burnham, and D. R. Anderson. 2001. Advanced features of program MARK. Pp. 368–377 *in* R. Field, R. J. Warren, H. Okarma, and P. R. Sievert (editors). Wildlife, land, and people: priorities for the 21st century. Proceedings of the Second International Wildlife Management Congress. The Wildlife Society, Bethesda, MD.

Wiens, J. D., R. G. Anthony, and E. D. Forsman. In press. Barred Owl occupancy surveys within the range of the Northern Spotted Owl. Journal of Wildlife Management.

Wildlife Society, The. 2008. Comments on the 2008 Fish and Wildlife Service final recovery plan for the Northern Spotted Owl. Letter to Director, Fish and Wildlife Service, 31 July 2008. <http://joomla.wildlife.org/documents/NSO_final_plan.pdf>.

Wingfield, J. C. 1985. Influences of weather on reproductive function in male Song Sparrows. Journal of Zoology 205:525–544.

Wingfield, J. C., M. C. Moore, and D. S. Farner. 1983. Endocrine responses to inclement weather in naturally breeding populations of White-crowned Sparrows (*Zonotrichia leucophrys pugetensis*). Auk 100:56–62.

INDEX

Endangered Species Act, 58
environmental covariates, 4, 5, 37

fecundity, 13–15, 17, 19, 20–28
  analysis of individual study areas, 13–14
  analytical methods, 13–15
  annual number of young fledged per territorial female
    (NYF), 13, 14
  annual time effects, 1, 2, 13, 20, 26–29
  a priori models, 13, 15
  autocorrelation issues, 14
  autoregressive time effect, 13, 20–22
  best fecundity models, 1, 20–22, 26–28
  competitive models in individual study areas, 2, 20–22
  competitive models in meta-analysis, 26–30
  definition of fecundity, 13
  effect of Barred Owls, 1, 2, 13, 21–22, 24, 26, 27
  effect of ecoregions (ECO), 2, 15, 17, 26, 27, 29–30
  effect of even–odd years (EO), 1, 2, 13, 14, 20–23
  effect of female age, 13, 20, 24
  effect of habitat, 1, 2, 17, 21–22, 24, 27, 29
  effect of land ownership (OWN), 2, 15, 17, 27
  effect of latitude (LAT), 2, 15, 17
  effect of precipitation, 2, 21–22, 25, 27
  effect of study area or group (g) in meta-analysis, 15
  effect of temperature, 2, 21–22, 25–26, 28
  effect of weather or climate covariates, 2, 26, 28
  estimates for individual study areas, 13, 20–26
  linear trends (T), 2, 13, 20–22, 27
  meta-analysis, 15, 26–28
  mixed models, 15
  nonparametric approach to estimate mean number
    of young fledged (NYF), 14
  normal distribution, 14
  normal regression model, 14
  quadratic trends, 13, 20–22
  random effect, 15
  residual variation, 14, 26, 29
  sex ratio at hatching, 13
  spatial variance among territories, 14, 26, 29
  study area–age class combination, 14
  temporal variance among years, 14, 26, 29
  truncated Poisson distribution, 13
  variance components analysis, 14, 26, 29
  weather effects, 2
federal lands, xii, 3, 4, 5, 8, 45, 53, 56, 58, 75, 76
field methods, 8
  band confirmation, 8
  banding owls, 8
  bands, 8
  brood patch detection, 8
  determination of nesting status, 8
  determination of number of young fledged (NYF), 8, 9
  fecundity estimates, 9, 13–15, 20–30
  locating nests, 9
  locating owls, 8
  noose pole, 8
  number of visits to each survey polygon, 8
  playback of owl calls, 8
  proportion of Spotted Owl territories occupied by
    Barred Owls, 9, 10

  protocol exceptions to reduce bias in fecundity
    estimates, 9
  snare pole, 8
  trapping methods, 8
  vocal imitations of owl calls, 8
forest management, 4

global models, 16, 17, 28, 45, 85
grand fir (*Abies grandis*), 8
Green Diamond Timber Company, xii, 6, 7, 58

habitat covariates, 10–12
  accuracy assessment of habitat maps, 10
  acronyms used in analysis, 10
  analyses of apparent survival, 10, 17
  analyses of fecundity, 10
  analyses of lambda (λ), 11
  analyses of recruitment, 10
  annual estimates of suitable owl habitat, 11
  baseline map, 10
  base map of suitable owl habitat, 10
  change detection, 10, 12
  criteria for defining study area boundaries, 10
  definition of suitable owl habitat, 10
  frame of reference, 10, 11
  percent cover of suitable habitat within study area, 10
  random effects, time specific, 11
  spatial scales of habitat covariates, 11
  time series of habitat maps, 10
  truncation of time series data, 11
Hoopa Tribe, xii, 5
human health, 4

incense cedar (*Calocedrus decurrens*), 8
information-theoretic methods, 1
insects, 8
interagency land management plan, 5, 58
Interagency Scientific Committee (ISC), 58

Kullback–Leibler information, 16, 17

land ownership categories, 13
latitude, 13
Lithocarpus densiflorus, 8
logging, 8
Louisiana Pacific Timber Company, xii

management, 3
Marbled Murrelet (*Brachyramphus marmoratus*), 3
mark-recapture studies, 3
mature forest, 8
maximum likelihood estimation, 16
median natal dispersal distance, 11
meta-analyses
  fecundity, 2, 15–17, 26–28
  lambda (λ), 9, 11, 13, 18–19, 43–56
  survival, 9, 11, 13, 17–18, 35–43
model selection
  AIC$_c$ model selection, 13, 16
  Akaike's Information Criterion, 16
  Akaike weights, 16, 17, 27, 29, 35–36

Composition: Macmillan Publishing Solutions
Text: 9.25/11.75 Scala
Display: Scala Sans, Scala Sans Caps
Printer and Binder: Thomson-Shore

1. Kessel, B., and D. D. Gibson. 1978.
   Status and Distribution of Alaska Birds.

2. Pitelka, F. A., editor. 1979.
   Shorebirds in Marine Environments.

3. Szaro, R. C., and R. P. Balda. 1979.
   *Bird Community Dynamics in a Ponderosa
   Pine Forest.*

4. DeSante, D. F., and D. G. Ainley. 1980.
   *The Avifauna of the South Farallon Islands,
   California.*

5. Mugaas, J. N., and J. R. King. 1981.
   *Annual Variation of Daily Energy Expenditure
   by the Black-billed Magpie: A Study of Thermal
   and Behavioral Energetics.*

6. Ralph, C. J., and J. M. Scott, editors. 1981.
   Estimating Numbers of Terrestrial Birds.

7. Price, F. E., and C. E. Bock. 1983.
   Population Ecology of the Dipper (Cinclus
   mexicanus) *in the Front Range of Colorado.*

8. Schreiber, R. W., editor. 1984.
   Tropical Seabird Biology.

9. Scott, J. M., S. Mountainspring,
   F. L. Ramsey, and C. B. Kepler. 1986.
   *Forest Bird Communities of the Hawaiian
   Islands: Their Dynamics, Ecology, and
   Conservation.*

10. Hand, J. L., W. E. Southern, and K. Vermeer,
    editors. 1987.
    Ecology and Behavior of Gulls.

11. Briggs, K. T., W. B. Tyler, D. B. Lewis,
    and D. R. Carlson. 1987.
    *Bird Communities at Sea off California:
    1975 to 1983.*

12. Jehl, J. R., Jr. 1988.
    *Biology of the Eared Grebe and Wilson's
    Phalarope in the Nonbreeding Season: A Study
    of Adaptations to Saline Lakes.*

13. Morrison, M. L., C. J. Ralph, J. Verner,
    and J. R. Jehl, Jr., editors. 1990.
    *Avian Foraging: Theory, Methodology, and
    Applications.*

14. Sealy, S. G., editor. 1990.
    Auks at Sea.

15. Jehl, J. R., Jr., and N. K. Johnson, editors. 1994.
    *A Century of Avifaunal Change in Western
    North America.*

16. Block, W. M., M. L. Morrison,
    and M. H. Reiser, editors. 1994.
    *The Northern Goshawk: Ecology and
    Management.*

17. Forsman, E. D., S. DeStefano, M. G. Raphael,
    and R. J. Gutiérrez, editors. 1996.
    Demography of the Northern Spotted Owl.

18. Morrison, M. L., L. S. Hall, S. K. Robinson,
    S. I. Rothstein, D. C. Hahn, and
    T. D. Rich, editors. 1999.
    *Research and Management of the Brown-
    headed Cowbird in Western Landscapes.*

19. Vickery, P. D., and J. R. Herkert, editors. 1999.
    *Ecology and Conservation of Grassland Birds
    of the Western Hemisphere.*

20. Moore, F. R., editor. 2000.
    *Stopover Ecology of Nearctic–Neotropical
    Landbird Migrants: Habitat Relations and
    Conservation Implications.*

21. Dunning, J. B., Jr., and J. C. Kilgo,
    editors. 2000.
    *Avian Research at the Savannah River Site:
    A Model for Integrating Basic Research and
    Long-Term Management.*

22. Scott, J. M., S. Conant, and C. van Riper, II,
    editors. 2001.
    *Evolution, Ecology, Conservation, and
    Management of Hawaiian Birds: A Vanishing
    Avifauna.*

23. Rising, J. D. 2001.
    *Geographic Variation in Size and Shape
    of Savannah Sparrows* (Passerculus
    sandwichensis).

24. Morton, M. L. 2002.
    *The Mountain White-crowned Sparrow:
    Migration and Reproduction at
    High Altitude.*

25. George, T. L., and D. S. Dobkin, editors. 2002. *Effects of Habitat Fragmentation on Birds in Western Landscapes: Contrasts with Paradigms from the Eastern United States.*

26. Sogge, M. K., B. E. Kus, S. J. Sferra, and M.J. Whitfield, editors. 2003. *Ecology and Conservation of the Willow Flycatcher.*

27. Shuford, W. D., and K. C. Molina, editors. 2004. *Ecology and Conservation of Birds of the Salton Sink: An Endangered Ecosystem.*

28. Carmen, W. J. 2004. *Noncooperative Breeding in the California Scrub-Jay.*

29. Ralph, C. J., and E. H. Dunn, editors. 2004. *Monitoring Bird Populations Using Mist Nets.*

30. Saab, V. A., and H. D. W. Powell, editors. 2005. *Fire and Avian Ecology in North America.*

31. Morrison, M. L., editor. 2006. *The Northern Goshawk: A Technical Assessment of Its Status, Ecology, and Management.*

32. Greenberg, R., J. E. Maldonado, S. Droege, and M. V. McDonald, editors. 2006. *Terrestrial Vertebrates of Tidal Marshes: Evolution, Ecology, and Conservation.*

33. Mason, J. W., G. J. McChesney, W. R. McIver, H. R. Carter, J. Y. Takekawa, R. T. Golightly, J. T. Ackerman, D. L. Orthmeyer, W. M. Perry, J. L. Yee, M. O. Pierson, and M. D. McCrary. 2007. *At-Sea Distribution and Abundance of Seabirds off Southern California: A 20-Year Comparison.*

34. Jones, S. L., and G. R. Geupel, editors. 2007. *Beyond Mayfield: Measurements of Nest-Survival Data.*

35. Spear, L. B., D. G. Ainley, and W. A. Walker. 2007. *Foraging Dynamics of Seabirds in the Eastern Tropical Pacific Ocean.*

36. Niles, L. J., H. P. Sitters, A. D. Dey, P. W. Atkinson, A. J. Baker, K. A. Bennett, R. Carmona, K. E. Clark, N. A. Clark, C. Espoz, P. M. González, B. A. Harrington, D. E. Hernández, K. S. Kalasz, R. G. Lathrop, R. N. Matus, C. D. T. Minton, R. I. G. Morrison, M. K. Peck, W. Pitts, R. A. Robinson, and I. L. Serrano. 2008. *Status of the Red Knot* (Calidris canutus rufa) *in the Western Hemisphere.*

37. Ruth, J. M., T. Brush, and D. J. Krueper, editors. 2008. *Birds of the US–Mexico Borderland: Distribution, Ecology, and Conservation.*

38. Knick, S. T., and J. W. Connelly, editors. 2011. *Greater Sage-Grouse: Ecology and Conservation of a Landscape Species and Its Habitats.*

39. Sandercock, B.K., K. Martin, and G. Segelbacher, editors. 2011. *Ecology, Conservation, and Management of Grouse.*

40. Forsman, E.D., et al. 2011. *Population Demography of Northern Spotted Owls.*